中国生活方式
A Real Beijing Dining Experience

健壹公馆菜
JE Mansion Cuisine

康健一 / 编

五洲传播出版社
CHINA INTERCONTINENTAL PRESS

序言

健壹之开端

王希富

中国的烹饪技术发展，从明清的宫廷菜开始走向汉族菜品的北菜品类，这与历代统治者居于北方特别是建都京城有着极为密切的关系。饮食与人的生活习惯、爱好，均会上行下效习与性成。当宫廷菜进入鼎盛时期，京城的八大楼与八大堂的顶级"庄馆菜"技术也已经炉火纯青。其实宫廷菜和庄馆菜的主要基础都是北方的鲁菜，只是前者服务于皇家，加入了满菜并融合了部分南菜，极其讲究材料和做工；后者服务于富豪、官宦、墨客与文人，当然也有庶民，最讲究色香味形和火候滋味。两者虽层次上下，但却各有千秋。因此宫廷菜和庄馆菜应当是京城甚至中国北方的优秀"头菜"品类。然而，近代中国百年的战乱纷扰和国事风雨，使泱泱大国之文化、技术等遭遇重创，富含烹饪技术的庄馆餐饮业也未能幸免。也有幸存的菜系艰难维系，但是上述两大系列菜类历经摧残，几近枝零叶败、花散根凋，甚至籽粒无存、难有传承了。当年宫廷御厨和八大楼的厨师如今已寥若晨星，即使有奇迹找到，却已是老迈年高、刀勺不动的活化石了。

　　说起以上闲话，拉杂无味，真正想说起的是健壹集团掌门人康健一先生的深厚志向与辉煌业绩。本来我自是一耕地出身的农夫，为名人大作写序实为佛头著粪，会笑坏大雅方家。可是学问无能，却是厨行世家，外祖父陈光寿祖上三代御厨，父亲、哥哥、舅舅有12人是八大堂或八大楼厨师。对于宫廷菜和八大楼庄馆菜，自幼便吃着、看着、学着、做着。那些技术、那些色香、那些味道和那些事儿，已经深入精神，渗入骨髓终生难忘。于此，或可为这部体大思精的作品说几句实在话，仅此而已。

　　说话人人都会，说实话却不容易。近几十年中国开始走上了复兴之路，经济发展，万民乐业。餐饮业和烹饪技术也随之兴业发展，这是百年来最值得庆幸的机遇。在京城，自身的老字号还没真正复活时，南方的粤菜、川菜便首当其冲打进京来，随后还有湘菜等等。这些名菜系本来就很硬气，又滨临海域，虽经百年沧桑也没太伤元气，近年又有很大发展和进步。本来历史上善吃北菜的京城食客，对"苟延残喘"的老庄馆菜的落败景象和无滋达味的口感已经腻味了，于是，就跟着当今的富豪、大户们吃起了新进京的"大菜"。京城的餐饮"旗舰"开始向南倾斜。请客吃北菜"羞于启齿"，请则龙虾鲍鱼，吃则川粤大菜，宁吃小馆子的"豆捞"，不吃老字号的"香糟"（指鲁菜的知名香糟菜，如糟熘三白等）。从晚清到民国，宫廷菜和庄馆菜炉火纯青的景象早已烟消云散、匿迹销声。这还不算悲哀，悲哀的是不少原来就是北菜的馆子，无论大小都开始"变卦"，要么搞"融合"，以借南菜的"仙气"，要么改口味，凡菜必加麻辣，凡烧必加鲍汁。就连旧时代非常有名气的八大楼之一，都卖起了粤菜。更不必说焦熘肥肠放辣椒、四喜丸子浇鲍汁了。问起来为何如此遭践北菜，曰：原来国营企业制度僵硬，闭门自守，人才流失，技术失传，菜品萎缩，每天三五件看家菜也变得面目皆非，无人问津。如今能明着挂着字号实则改换门庭的还能活着，那些连如此能耐也没有的，就只有把字号卖给私人，职工解散回家了。

本来事物发展总是强者生，弱者亡。但是，京城曾经是孕育和发展北菜的母亲和摇篮，不少仁人志士心中酸楚，难忍静观不语。然而，此事该谁来管，谁能够在泥沙混杂的大浪中力挽狂澜，拯救即将沉沦的北菜？

此时，京城异常活跃的餐饮行业中忽然出现了一个"宅门菜"，让京城食客眼前一亮。康健一先生便是"宅门菜"的创始人之一。健一先生与我相识时，他已经有不止一个宅门酒家了，而且经营都很火爆。我答应做他的顾问便去他的宅门"看样"，想知道"宅门菜"有无提高成高档北菜的可能。看了这种创新经营的店面，品尝了其中的菜品，确实很有感受。一是殿堂环境的装修，开始出现中国传统建筑要素，为顾客打造了一种回归老宅的感觉；二是出现了中低档北菜供应。在满墙燕窝鲍翅满眼海鲜的红灯绿彩中，京城食客终于看到了一点北菜的踪影，尝到了一丝老年间庄馆菜的滋味，坐上了一回榆木擦漆的扶手椅，有一种旧地重游、旧福重享的感觉，仅此，已经欣喜至极了。这或许就是新兴大宅门的成功。

健一先生不主故常，总在思考革故鼎新。他不满足大宅门的成就，也看出了其中尚有不足，甚至硬伤所在。我于是无可讳言实话实说，他也能虚己以听。一是室内装修过于繁琐而缺少精雅，不够大气磅礴；二是菜品过度融合也显杂乱无序，难上档次。吃是能吃到不少北菜，其实真说滋味还比不了当年的"二荤铺"。就连切面铺的面条种类、清汤卤、浑汤卤、素卤、炸酱、浇头氽的多种做法，还不太清楚，何言恢复北菜和宫廷大菜。就当时的情况，大宅门以保持原状提高水平为好。想打造高水平的庄馆菜只有重打鼓另开张。这是我和健一先生的共识。其实健一先生是早有此想法，并且投巨资建造了一个新型模式的酒店，

当时健一先生提起"公馆"，其实他所言之公馆已不是古代诸侯的离宫别馆，而是民国时期官宦、军界政要们的豪宅。此时京城也正在流行"公馆"，而且所传

神秘异常，如不露真字号，或"无名氏画像"，或"编号大院"。内有离宫别馆，物资特点供应，餐具是"红色官窑"，厨师身怀绝技。于是不少外地名流阔佬、艺人食客趋之若鹜。在京城奔走寻觅，结果"公馆"是有，除了价格惊人之外，不少只是普普通通的买卖，不过是到处都有的吃食而已，有名无实。

历史的沧桑让人们早已把民国时期的公馆遗忘。而且，一般人的理解，那些上层社会尔虞我诈的场所，军政混杂的场面，灯红酒绿的奢华，如痴如梦的生活，如今已没什么人再想去回味了。旧公馆还有精华吗？

凡事之思索，必须求其内涵，其实宫廷菜就是帝王无穷的奢华，今人只有唾弃而已。我们还研究何用？但是，古为今用，必去其糟粕，取其精华。健一先生提出此点切中了"新公馆"的要害。

我曾经就这样的问题，请教过我的大哥希荣，他在上世纪40年代被军界从八大楼挖到公馆做主厨。他的见解有三。一是当时的公馆之主皆是军界要员，所设菜系京津地区皆为鲁菜，要求比八大楼更为严格，因为所来之宾客经常是要员的上司，饭菜不可稍有马虎，不然，轻者影响关系，重者丢官罢职，厨师中绝无滥竽充数；二是管理严格，制度皆为军纪，服务之周到所有饭铺都无法相比；三是场地设施家具装饰要有极强的文化内涵，绝不能出现"关公战秦琼"的笑话。仅此三点，就可以归纳出旧公馆的精华所在。秉承这样的原则，将之有机组成今日之"公馆"，让其为新时代所用，不失为新发展的途径。

果然，京城出现了如此推陈出新的健壹公馆，继而又有了玉泉山下的健壹景园。

健壹集团旗下健壹公馆、健壹景园的建筑，极具民族特色，这是其本质所决定的。

健壹公馆因为是京城地区的中式公馆，其规格没有使用红墙黄瓦的宫殿式，便躲开了"一眼假"的暴发户风格。从东四环路辅路南行，便可见别致却不张扬的馆门，驱车进入，沿布满草坪灯的小路，曲径通幽前行，豁然开朗，便看到筒瓦黑活的歇山建筑大门，翘角重檐，门前有大气磅礴的抱厦，可以在风雨中迎接宾客。门前的碧水、曲栏围衬着主楼，主次分明，相得益彰。门后有大气的筒瓦歇山式抱厦，衔接主楼。主楼则是简约平素的西式风格，加入部分砖雕、石雕的装饰，使整体造型成为中西合璧的"民国式"建筑，很有老式公馆的风格。进入楼门，便可见宽阔的厅堂，仰视天花，藻井高悬，金龙下视。宽阔的楼梯，别致的甬道，可以一步一景地看到各类布置得体的家具、书画、瓷器和艺术品，沉稳和谐。北式家具一律榆木擦漆，雅而不俗；夹杂其中，尚有雕饰繁缛的香案类便为南式金漆。但没有使用硬木三色，这便是健壹公馆室内陈设的高明之处——明快而简约。为顾客创造了一个极具文化特色的环境和宽阔、明净、和谐的使用布局。因而，是地道的公馆而不是虚假的皇宫。

坐落于玉泉山脚下的健壹景园则仿佛是矗立在时间之外的桃园之乡，庭院之中草木隐约，如水墨画中的渲染，体现出古代中国文人推崇的低调含蓄之美。中式风格贯穿整个庭院，整座搬运而来的江南老宅，完整的传统四合院，灰色砖墙和从民间收藏而来带有浓厚历史痕迹的镂空花纹棕色木门、刻有名士贤语的匾额楹联，旧时宫廷记忆，在此找到了新生命。

菜系决定了健壹集团的灵魂和主旨。既然在京城，就下决心拿出北菜的精华：宫廷菜和八大楼的庄馆菜。开业前，集团提前对厨师进行了培训，经过漫长的传授、学习和上灶实践，冷荤、热炒、汤羹、烧烤、甜品、粥品、饽饽，已近百品，全部是原汁原味的老菜。在补充了其他所需要的融合菜系，豪华的燕翅席、参翅席，乃至满汉全席的半桌、一角，大厨们已经得心应手了。一经推出，即赢得满堂喝彩！

服务更不必说，集团专门聘请名家管理，从经营、销售、服务和一切细节末微都难有纰漏，追求完美。使公馆没有"洋酒店"的"道貌岸然"，既有宾至如归的温馨，又有文化大家的气派。

正是：洗尽铅华归真谛，范饮模餐静里商。

Preface

The Origin of JE Group

Wang Xifu

After the end of feudal China, the development of Chinese culinary techniques made a transition from Manchu-influenced royal cuisine toward Han-style "northern cuisine". This change in direction was closely related to the fact that for several dynasties all of the emperors had resided in Beijing, and many of their eating habits were passed on from one generation to the next. By the time royal cuisine had reached its peak of popularity, the dishes made by the "Eight Great Restaurants of Beijing" were prepared using extremely refined techniques. Royal cuisine and zhuangguan cuisine (literally "farm and estate") both share a common ancestor, namely Lu cuisine of Shandong Province. The main differences between the two are that royal cuisine was served to the royal family, and thus absorbed some influences from Manchu and southern cuisine, with heavy emphasis on the quality of the ingredients and workmanship. Zhuangguan cuisine was mostly enjoyed by wealthy lords, officials and scholars, and to some extent the common people, the main focuses being the combination of "color, aroma, taste and appearance", as well as the degree and duration of heating.

Although royal and zhuangguan dishes each have their own unique characteristics, they are essentially two different levels of the same type of cuisine, thus they are often collectively known as the representative cuisine of Beijing, or sometimes even China. However, contemporary China has seen its share of wars and turmoil, which have influenced its culture, including culinary culture and techniques. Royal and zhuangguan dishes in particular have become "endangered species" among China's widely varied system of local cuisine, with the recipes for many dishes having disappeared completely or being passed on by only a handful of people. Most of the direct disciples of chefs from the end of the Qing Dynasty (1644-1911) are so aged that they cannot completely recollect the recipes, which have never been written down in full form.

But of course there's no point simply reminiscing about the past. What I really want to talk about in this foreword are the admirable ambitions and outstanding achievements made by Mr. Kang Jianyi, founder of JE Group. I grew up on a humble farm and am not highly educated, so I may not be worthy of writing a foreword for a respectable man such as Mr. Kang, but there is something quite special about my family: My grandfather Chen Guangshou and the three generations before him were all royal chefs in the Forbidden City, and another 12 of my other family members were chefs at the Eight Great Restaurants of Beijing, including my father, brother and uncle. I was taught to make royal dishes from childhood, and the tastes and techniques involved in them will follow me to the end of my days. For the introduction to this book of massive in scope, I would like to share a few honest words.

Speaking is easy, but speaking the truth is often difficult. Over the past several decades China has begun a sort of renaissance, resulting in rapid economic development and a great improvement in the standard of living. At the same time, our catering industry and culinary techniques have also seen much progress, and Chinese cuisine has never been so complete and accessible. In Beijing, before the classic restaurants had undergone full reincarnation, southern styles had already been booming in the capital, led by the cuisine of Guangdong and Sichuan, followed soon by that of Hunan, and so on. Due to their locations, these styles were not heavily affected by war or other historical events. After the end of the Qing Dynasty, residents of Beijing longed for something different from the local cuisine, and these southern dishes became immensely popular, thus the major restaurants of the capital began catering to their customers' demands. Treating guests to northern dishes even became an embarrassment, whereas a meal consisting of Cantonese or Sichuanese dishes, such as lobster and abalone, was considered much more high-class. Diners would prefer to eat at a small hotpot restaurant than be caught dead eating Lu dishes. To make matters even worse, northern restaurants of all sizes eventually began "adapting" their menus, with either north-south "fusion dishes" or adding southern flavors and ingredients to make their dishes more "classy". In fact, one of the Eight Great Restaurants of Beijing even transformed itself into a Cantonese restaurant, adding Sichuan peppers to their fried sausages and abalone sauce to their four-joy meatballs. As for the reason why northern cuisine fell to such a low status, it was a result of the fact that local restaurants had a very rigid way of doing things; they were closed to new ideas, thus many talented chefs left, taking with them their unique skills, and the handful of remaining specialty dishes soon lost the interest of diners. Under these conditions, none of the classic restaurants could keep their original name but change their style of cooking to survive, thus many chose to sell the name of their restaurant to a private owner and cash in while they could, leading to the end of their time-old traditions.

The survival of the fittest rule has always applied to the restaurant industry. But were the people of Beijing, the cradle and home of northern cuisine, able to simply stand by and watch their culinary culture disappear forever? Of course, many felt opposed to transition, but there was not much they could do, as there was no one to held responsibility for the survival or extinction of their local dishes.

Fortunately, in recent years Beijing has seen a revival of its traditional cuisine culture, known as "mansion cuisine", and Mr. Kang Jianyi is one of the forefathers of this movement. When Mr. Kang and I first met he already owned several mansion cuisine restaurants, all of which were immensely popular. I agreed to be his adviser and "overseer" of his mansion dishes, wondering if his food would be able to heighten the status of northern Chinese cuisine. Upon visiting one of his innovatively designed restaurants, I admitted that I was quite impressed. The establishment itself was constructed and decorated in the traditional Chinese style, giving one a sense of entering an old Chinese mansion, but I was also surprised to see that they served mid- and low-range northern dishes. Among all the sea cucumbers and shark fin that have overtaken upper-class Chinese cuisine, locals and visitors to Beijing now finally have a chance to enjoy real traditional northern Chinese food, to taste these great dishes that have been lost for decades or even longer, not just read about them in books or see them in museums. This has given me hope and brought me much joy, as it's possible that mansion cuisine is solely responsible for the revival in popularity of true northern Chinese food.

Mr. Kang is a leader, not a follower. He is not satisfied with his Dazhaimen Hotel, seeing many faults and failures in its design and operation. I can be straightforward with him about anything, and he'll always listen with an open mind. I've told him that the furnishing of Dazhaimen is overly complex and lacks refinement, thus influencing its overall atmosphere, and that the dishes are too fused and varied, making it impossible for them to fulfill the desires of more demanding diners. Although the hotel does serve many northern dishes, none of them come close to those made by the restaurants of the olden days, and even simple things like the different types of noodle sauces are not up to standard, so it would be difficult for dishes like these to singlehandedly cause a comeback in northern and royal cuisine. Dazhaimen has already come a long way since then, improving their level of cooking skill while retaining their original direction. But I told Mr. Kang that if he really wanted to serve the best zhuangguan dishes, he would have to start from scratch. As it turns out, Mr. Kang had also thought this way for quite some time, so he decided to invest a vast amount of funds in the construction of a new type of hotel.

He called this style of establishment a "mansion", which he explained as not being the kind that lords of feudal China stayed in when they were away from their palaces, but rather the large estates that officials and military leaders during the Republican Period (1912-1949) had built using government funding. During this era the term "mansion", due to its sense of mystery and ambiguity, was popular throughout Beijing as a type of high-class restaurant featuring expert chefs and fancy dining ware. These establishments attracted wealthy diners from throughout the country, and eventually were so commonplace that the capital's resources became greatly exploited, and many of these so-called "mansions" were simply relatively average restaurants with luxurious settings and outrageous prices.

Time led to the inevitable collapse of the mansions of the Republican Period, leaving behind impressions of places featuring nothing but deception, corruption and excessive extravagance, thereby tainting the status of these restaurants as places that no one would want to visit, let alone establishment and operation.

But Mr. Kang believes that mansions do have their place in the history and culture of China; royal cuisine was a reflection of the vast riches possessed by the emperor, and although by practical standards we can only dream of spending so much money on a single meal (the emperor would be served 200 dishes and eat from only a dozen or so), royal cuisine culture can still be represented in a less extravagant form. So, Mr. Kang proposed the concept of the "new mansion", which could represent the positive aspects of royal and mansion dishes, without all the excessiveness.

My eldest brother Xirong is an expert on the mansions of the olden days. In the 1940s the army relocated him from his job at one of the Eight Great Restaurants to a mansion restaurant, where they had him act as head chef. He has concluded that these establishments follow three main principles. The first is that the main clientele at mansion restaurants are military personnel, and the food they serve is all Lu cuisine, with stricter adherence to tradition of the

Eight Great Restaurants; since many of the guests are of even higher official status than the hosts, if there are any shortcomings in the meal the host may lose his status as an official, an outcome no chef would ever want to encounter. Second, the management system is also very strict, organized much like a military organization, and the service must be just as impeccable as the food. Third, the furniture and decoration of the dining rooms must comply with cultural standards, with no room for errors involving the incorrect mixing and matching of items from different eras, which would be joked about as a sort of "Qin Shi Huang battling with Genghis Khan". These three principles are the essence of the mansions, and are worthy of being transmitted to the new style of mansion.

After Mr. Kang built JE Mansion, based on this concept of applying old principles to new ways of operation, he continued his endeavor by establishing JE Garden, located at the foot of Beijing's Yuquan Mountain.

Both JE Mansion and Garden feature characteristically traditional Chinese architecture, which is a result of the ideas that they are based on.

For JE Mansion, Mr. Kang decided against the "fancy but fake" design, avoiding the use of the red brick walls and yellow tiled roofs seen in the palaces of the ancient capital. Easily reached within close proximity of the Fourth Ring Road, the entrance to JE Mansion is very simple and modest. After entering the gate, you will pass down a small path lined by grass and lampposts, before coming to the main building, which bears a roof with pointed corners, layered eaves and semicircular tiles. Stepping out of the vehicle under the extended canopy, a short path bordered on either side by small ponds, and separated from the walkway by rustic handrails, leads you to the main entrance. The inner annex is large and spacious, featuring spider web ceilings from which golden dragons gaze downward, broad open stairways, unconventional walkways, and varied yet harmonious collections of antiques, including wooden furniture, old books, porcelain vases, and fine paintings. The northern-style furniture are all constructed from elmwood and lightly glazed, for an appearance which is sophisticated yet humble, unlike the shiny glossed furniture of the south. But what is most surprising about the furnishing of JE Mansion is the fact that although there is very little variation in the colors used, as a whole it is still very attractive, as its design is simple and direct. Not only is the layout imbued with a sense of Chinese tradition, it is also very spacious and practical. It is what Mr. Kang calls a "real mansion", but not a "fake palace".

JE Garden follows similar notions, its location at the base of Yuquan Mountain resembling a setting out of a fairytale, full of lush, picturesque scenery, acting as a worthy representation of the modest and reserved beauty so highly regarded by the literati of ancient China. The very traditional buildings on the premises include complete quadrangle courtyards surrounded by gray brick walls, and inside the rooms antique objects such as etched wooden doors and inscribed tablets, collected by Mr. Kang from throughout China, have found a new home where they may share their stories of the past.

The cuisine of JE Group has determined its essence and significance. Being located in Beijing, Mr. Kang is dedicated to providing royal and zhuangguan cuisine which can truly represent the traditional dining culture of the ancient capital. Before JE Mansion's official opening, all of the chefs underwent rigorous training on the food they would be serving, including appetizers, fried dishes, soups, baked and roasted dishes, congees, pastries and desserts, for a total of about a hundred authentic traditional recipes. They also received additional training for other types of meal settings, such as fusion-style meals, extravagant banquets, and even partial Manchu-Han Imperial Feasts. By the end of the training program, the team of chefs had honed a set of skills with which no other restaurant could compare. Beginning with their first official meal, the JE Mansion kitchen team has given customers an unforgettable dining experience.

What Mr. Kang strives to achieve with both his cuisine and series of hotels is a departure from luxurious falsehood, and a return to unadorned authenticity.

前言

以文化的积淀诠释经典
—— 写在《健壹公馆菜》出版的时候

边 疆

一个涉及文化、技艺、业态、品牌的概念——官府菜，被业内专家、学者、老板乃至厨师讨论和争辩多时。而康健一先生却在默默无闻的十年里，在全国打造了五家公馆型酒店，研发了近百道公馆菜点，当他的探索集约在《健壹公馆菜》一书且即将公开向国内外出版发行时，我们不得不赞叹他以深厚的多样文化艺术的积淀所诠释的官府菜经典。

五千年的中国历史，造就了深厚的中华文明，作为以食为天的华夏子孙所创造的内涵丰富的饮食文化亦成为文明历史百花园中的奇葩，不仅被孙中山、毛泽东等伟人尊为"对世界的一大贡献"，也被历代后人不断续写。

官府菜作为采集民间美食技艺并为专业厨师加工精制、为美食家丰富完善的美食文化，经历代不断补充扬弃，成为中国饮食文化的一大结晶。据考，八大菜系流传下来的许多菜点都是经各地官府采集、加工、提高后在民间流传不衰的。

四川旅游学院教授、饮食文化专家杜莉在《中国饮食文化》一书中说道："官府菜，亦称公馆菜，是封建社会官宦人家制作并食用的肴馔。官府菜注重摄生，讲求清洁，工艺上有独到之处，不少家传美撰遐迩闻名。对于官府菜，《晋书》有'庖膳穷水陆之珍'的记载，唐人房玄龄有'芳饪标奇'的评语，可以说，官府菜是封建社会达官显贵穷奢极侈、饮食生活争气斗富的历史见证"。专家是如此定义官府菜的，同时也阐述了官府菜在多年处于封建官本位的中国历史条件下的生成背景，及其在中国饮食文化中的重要地位。

官府菜成为中国饮食文化之奇葩，还在于它采诘于不同地区、不同时代，又根据不同生活背景和不同爱好的权贵意志改造，经过不同菜系的高厨加工制作，流传下来就形成了不拘一格而追求极致的精品。因此，能够掌握官府菜精髓的人应该不仅是模仿，更是有能力采百家之长，于传袭中彰显自我。康健一先生是我以为的官府菜研究和实践专家中少有的佼佼者。

爱好国画、古玩；搞过收藏、拍卖；钻研古建、装潢……康健一对历史文化有着深厚的积累，多年研究古典文化培养了他的严谨习惯，而对艺术的酷爱也使他对美的追求不乏浪漫，这些一旦遇到餐饮和酒店这个将美全方位示人的行业，康健一的运营就完全不同于业内人士了，他要将文化、艺术与现实生活完美结合，"做一个让中国人骄傲，让外国人尊重的酒店"。走进健壹公馆，几乎每个人都可以找到这个感觉。

仅有感觉还不能全面认知健壹，或者说还不足以表现康健一先生的思维。从庭院的风水规格与视野景观，到建筑的雕梁画栋与中世纪花纹；从装潢的皇家帐幔与王室雕塑，到家具的明清卧榻与欧陆靠椅；从店堂的古筝弹奏到员工的时尚装束，都无不全面体贴地在你身边将公馆文化娓娓道来。

十年的打造，康健一最大的收获之一是培育起独特的餐饮文化体系，从而构成他的健壹集团产品的核心竞争力。健壹的精神文化就是力求"把事情做到极致、做到独一无二"的理念；其行为文化就是在精品价值观指导下的中式理念结合西式的人性化服务；其物质文化则是以传统中国宫廷菜点为基础，融汇现代理念和时尚技法的官府菜，康健一定名为公馆菜体系，这一体系成为健壹集团经营不断发展的强大引擎。

公馆菜是一个整合的概念，纵观各家公馆菜的发展，差异与特色是共同的特点，研究和发展公馆菜则应遵循这个理念，不然，公馆菜就将会因趋同而逐渐消亡。康健一先生从建设公馆开始就将这个理念贯穿于菜品开发的始终，他邀请到公司的几位菜品顾问就表现了他的这个理念。一位是著名古建专家、宫廷菜传人王希富先生。王老的外祖父为清光绪皇帝御厨，父亲与兄长多为厨师，王老11岁就会做宫廷菜，他收集和掌握了近800道宫廷菜的菜谱和制法，对清末皇家饮食礼仪和饮食典故了如指掌，是京城屈指可数的宫廷菜传人。还有一位来自台湾的任全潍先生。任先生初入行业时学习日餐烹饪技术，之后曾在法国留学并长期工作，多年在法国米其林三星餐厅担任大厨。长期的专业训练和工作经验，使他成为融汇了中餐、日餐和法餐各种技法和经典的大师。而且无论是正餐大菜还是餐前餐后的冷菜、甜点，甚至是基本的装饰花样，都有自己的独到的见解和创新，而且对于出品品质更是精益求精。王老先生和任先生之所以分别成为健壹的顾问和技术总监是因为康健一先生认为，他们都是能把事情做到极致的人。

传统与时尚的结合，形成健壹公馆菜最大的特色。从健壹公馆菜中我们看到了中国宫廷菜的技艺功底，也看到了欧陆菜的重视食材，还看到了台湾菜的口味精致；看到了中餐的多重技艺与调味，也看到了西餐的简洁制作与造型。一个以康健一先生古典文化积淀和现代艺术修养为审美水准的公馆菜，集华贵、经典、悦目、美味为一体，成为当代公馆菜的一个代表，也成为今日特色、高档、精致美食的最好诠释。

中国烹饪大师，知名餐饮职业教育家李刚先生曾将发展烹饪文化总结为烹饪学的最高境界，即"具有健康的烹饪思想和科学的烹饪发展观，能传承发扬优秀的烹饪成果，将历史发展积淀下来的精神财富、物质财富和现代文明成果有机结合，赋予她新的生命，供人类享用。"用李刚先生的定义介绍健壹公馆菜之精髓，是再恰当不过的。

随着社会经济的发展，人们生活水准的提升，旧时的官府或公馆的生活条件早已比比皆是，而生活的社会化减少或缺失了公馆菜生成的条件。日益增长的需求和逐渐失传的美味烦扰着我们，万分庆幸，还有康健一先生这样执着的公馆菜守护者，用他精致的服务理念和融汇态度续写着华彩乐章，用他深厚的文化底蕴和创新精神传承着美艳经典。

祝《健壹公馆菜》早日出版，以给我们带来对中国美食文化更多的敬畏。

二〇一三年八月八日
于北京龙湖花盛香堤

JE Mansion Cuisine

Foreword

Reflecting Culture through Cuisine

Written during the publication of *JE Mansion Cuisine*

Bian Jiang

"Mansion cuisine", a concept involving culture, culinary techniques and business operations, has been discussed and debated by industry experts, scholars, restaurant owners and chefs for many years. But, over the past decade, Mr. Kang Jianyi has quietly established five mansion-style restaurants throughout China, which serve almost 100 mansion dishes researched and developed by Mr. Kang himself. His explorations are collected in *JE Mansion Cuisine*, which is being prepared for publication at home and abroad. Indeed, Mr. Kang's achievements in harmonizing Chinese culture and history together with the classic mansion cuisine recipes are quite inspiring.

With its 5,000 years of history, China possesses a very rich cultural background, including a culinary culture that is among the most varied and highly developed of the world. Both Dr. Sun Yat-sen (1866-1925) and Mao Zedong (1893-1976) said that Chinese cuisine is "a great contribution to the world". The Chinese cuisine is still under promotion by contemporary people.

Mansion cuisine, as a representative collection of Chinese folk culinary methods, as well as a cuisine culture featuring finely-polished cooking techniques and an abundant variety of types of dishes, has been modified throughout late feudal and early contemporary China, earning its place in

Chinese culinary culture. According to historical sources, many dishes in the "eight styles of Chinese cuisine" were collected from the menus of mansions throughout China, then processed and improved, before taking on their current appearance.

Du Li, Cuisine culture expert and professor of Sichuan Tourism University, wrote in her book *The Cuisine Culture of China*: "Mansion cuisine refers to delicacies prepared and enjoyed at the homes of officials in feudal China. It focuses on healthiness, emphasizes hygiene, and possesses a set of unique production skills and a large number of family recipes. Mansion cuisine was mentioned as early as Tang Dynasty (618-907) in the *Book of Jin*, in which chancellor Fang Xuanling (579-648) described mansion cuisine as being able to satisfy the extravagant standards of officials at the time." This became the definition of mansion cuisine among experts, leading to its classification as a type of cuisine enjoyed by officials of feudal China, as well as its important historical status in Chinese cuisine culture.

Mansion cuisine as a unique element of Chinese cuisine culture also stems from the fact that the dishes were collected from different regions and eras, adapted by people of different backgrounds and objectives, and prepared by chefs specializing in a range of cooking styles, thus after several centuries formed a set of varied and refined delicacies. Therefore, if someone were to truly grasp mansion cuisine, then simple imitation would not suffice, it would have to be someone who is willing and able to learn from other people and regions, and at the same time add their own unique perspective. I believe Mr. Kang Jianyi is one of the few people worthy of being referred to as an expert on mansion cuisine.

Mr. Kang loves Chinese painting and antiques. He has worked in collecting and auctioning industry, and intensively studied ancient architecture and décor. He has accumulated a vast expanse of knowledge related to Chinese history and culture, years of researching classical culture have brought him rigorous habits, and his obsessive love of art has driven his endless pursuit of beauty. When these characteristics are combined with the hotel and catering industry, which serve to provide luxury and enjoyment, Mr. Kang's style of operation is certain to be completely different from anyone else's. He wishes to create a flawless combination of culture and practical living, and establish "a hotel which can make the Chinese proud, and obtain foreigners' respect". When stepping into JE Mansion, this is very likely what you will feel.

But this feeling alone is not enough to fully understand JE Mansion or Mr. Kang's thoughts. From the layout and scenery of the complex's courtyard, to the ornate rafters of the buildings; from the royal-style curtains and sculptures, to the combination of imperial-age Chinese and modern European furniture; and from the guzheng player in the main hall to the unique design of the staff's uniforms, every detail of everything reflects the mansion culture.

After ten years' endeavor, Mr. Kang has created a unique cuisine culture establishment, which is one of his greatest achievements and constitutes the core competence of JE Group. The ideology of JE Group is striving for exceptionality and absolute uniqueness. The group's philosophy of service combines traditional Chinese concepts with western practical solutions. And its cooking methodology is based on royal Chinese dishes, fused with modern ideas and the latest cooking methods. This entire system of concepts makes up what Mr. Kang refers to as mansion cuisine, which JE Group is dedicated to continuously developing and exploring.

Mansion cuisine is an integrated concept, involving a wide variety of locations and styles, thus inconsistency and variation are its distinguishing characteristics. Anyone who researches and develops mansion cuisine must follow this pattern; otherwise, the mansion cuisine will fade away if all types of mansion cuisine turn to be the same. Mr. Kang has kept this concept in mind from the very beginning, and his cuisine advisers have ensured that this is fulfilled.

One of these advisers is ancient architecture expert and royal Chinese cuisine successor Mr. Wang Xifu. Mr. Wang's grandfather was a royal chef for the Guangxu Emperor, and his father and eldest brother are all chefs. Mr. Wang learned to cook his first royal dish at the age of 11, and has collected almost 800 royal recipes. Not only can he prepare most of these dishes and describe their background stories, he is also extremely familiar with late-Qing Dynasty royal cuisine etiquette, making him one of the few true successors of Beijing's royal cuisine.

One of the other advisers is Mr. Ren Quanwei, from Taiwan's "3/2 Chateau", which is currently the only completely private restaurant with only five small dining rooms in Taiwan. Mr. Ren lived and worked in France for many years, as a chef at a Michelin three-starred restaurant. In Taiwan he also owns western restaurants, bakeries and pastry shops. Mr. Kang invites Ren as adviser of JE mansion,

because he believes Ren is also a person striving for exceptionality.

The combination of tradition and fashion has become a trademark of JE Mansion's cuisine, including techniques straight out of the royal Chinese kitchen, the European emphasis on ingredients and simplicity of preparation and presentation, as well as the attention to detail of Taiwanese cuisine. The cuisine of JE Mansion, fulfilling the high expectations of Mr. Kang's cultural and artistic background, includes dishes which are luxurious, classic and pleasing to all the senses, serving as the most ideal reflection of China's new style of high-class dining.

Renowned chef and culinary educator Mr. Li Gang once referred to cuisine culture's developmental stage as the highest attainment of the culinary arts, saying that "Having a health-based and scientific outlook toward cuisine can lead to outstanding achievements in the spreading and passing on of the culinary arts, allowing spiritual and material wealth to unite with the achievements of modern civilization, thus allowing these creations of mankind to receive new life, and be enjoyed by mankind once again." This definition of cuisine culture is a flawless representation of JE Mansion's cuisine.

With the economic development of society, people's living standards have been greatly increased. Although the living standard of the mansions and manors of olden times has been achievable, our social habits and customs have weakened or lost some of the conditions which allowed for mansion cuisine to be created. Our incessantly increasing desires and constantly disappearing traditions make it difficult for mansion cuisine to survive, but fortunately Mr. Kang Jianyi is dedicated to preserving and promoting mansion dishes, writing the next page in the history of Chinese cuisine culture. JE Mansion's exquisite service concept, integrated creative perspective and deep cultural background ensure its status as the forerunner of new Chinese cuisine.

I look forward to the publication of *JE Mansion Cuisine*, and I hope that more and more people will gain a deeper understanding of Chinese cuisine culture through this book.

Aug. 8, 2013
Dragon Lake, Longtan Park, Beijing

概说宫廷菜

熊 蕾

从三皇五帝开始，中国历史上有诸多朝代诸多宫廷，自然也有诸多宫廷菜。但是如今所说的宫廷菜，一般都是指清朝宫廷皇家膳房的菜点。作为中国最后一个封建王朝，也作为由中国东北一个少数民族入主中原的王朝，清朝宫廷菜继承了明朝及以往朝代的宫廷菜特点，也吸收了蒙、回、满等不同民族的风味膳食，可以说集历朝历代宫廷菜和民间高档菜之大成。

宫廷菜的基础

主宰整个清朝宫廷菜的，是鲁菜。产生于滨海的山东省的鲁菜，是代表中国烹饪艺术的八大菜系之一。山东是崇尚美食的中国儒家思想创始人孔子的故乡，烹饪艺术源远流长。鲁菜作为一个成熟菜系成形于元代，但是其历史足可追溯到3000多年前的春秋时期。山东既临海，又拥有物产丰饶的平原和山区，盛产海产品、海盐、五谷杂粮和各种农产品，山东人民自古就开发出各种调味品，为鲁菜系的形成奠定了厚实的物质基础。

孔府老宅

由于多少世纪来的战乱和饥荒迫使很多山东人背井离乡，山东菜也随着离开了自己故乡的人到了他们的新家，在中国北方地区，包括北京、天津以及东北等地流行开来。有一些闯关东到了东北的山东老乡，在满清入主中原之前，就成了满清皇家的御厨。

清朝给皇上做饭的厨子都是汉人。满人不会用满人来做饭。因为在当时做饭被认为是低人好几等的职业。满清坐了天下以后，所有的满人都是坐享其成的，不用干活，全吃俸禄。满人能去当低人几等的厨子去么？这是绝不可能的。

不仅清朝的宫廷菜以鲁菜为主。据说，清朝第三个皇帝，也是满清入主中原的第一个皇帝顺治（1368-1661）和满清皇家在1644年离开他们在东北的根基进驻北京的时候，紫禁城里还剩下七个给刚被推翻的明朝皇帝做饭的御厨。这七个前明御厨全都是山东人。这说明在汉人坐天下的明朝，宫廷菜也是以鲁菜为主。

所以，清朝的宫廷菜是以鲁菜为基础，同时吸收并提高了一些原来满族人在东北吃的满菜。满菜大部分都是用东北的原料做的菜，比如用鹿做的菜，用熊掌做的菜，用飞龙做的菜，像烤鹿方，鹿筋狍子肉，都是真正的东北菜，也是典型的"老"宫廷菜。

典型的满菜还包括满洲饽饽，也就是点心，这也成为宫廷菜的一个重要组成部分。满族人原来在东北的时候是游猎民族，但他们以能够做各种点心为自豪。他们举行盛会的时候，富裕的部族争相摆出饽饽桌子，上面码放着一层一层的点心，谁的饽饽桌子摆得高，就代表谁有钱。可惜满洲点心现在传下来的没有多少了，但是传下来的几款至今仍很受欢迎，比如萨其马，芙蓉糕，奶卷子等。

不惜工本

宫廷菜以鲁菜为基础，但是更精细，可以说"不惜工本"。如果北京最好的庄馆菜做一斤肉，用二斤原料，那么宫廷菜做一斤肉要用八斤原料。得选最好的，最贵的，并剔去大部分"等外料"。给皇上做的万字扣肉，必须是八层肥八层瘦，而瘦的必须是肥的两倍厚。要找这样一块合适的肉，可能要杀十几头猪，而且还要"人工制造几层五花"才得可用。

再比如做灌汤黄鱼，一条鱼买来就得1000多块钱。鱼肚子里是燕窝鱼翅，鲍鱼海参。还要吊好汤，做好味儿。外头还有好多材料，都要求最好的。光原料就得2000多块钱。做工又费事，蒸了煮，煮了蒸，再炸，来回不知折腾多少遍，才能成形。这个过程中，鱼还不能变形，破一个，都得算钱。这样一条灌汤黄鱼的成本不下5000块钱。所以，现在好多餐馆做不起宫廷菜。即使少数几家如健壹公馆能做宫廷菜，像灌汤黄鱼这样的菜品也赚不了多少钱，就因为成本太高。

宫廷菜还应该讲究的是餐具，专家认为这是宫廷餐饮文化一个必不可少的组成部分，即所谓"美食美器"。专为一道菜烧制一套餐具的情况，并不罕见。

清代宫廷对炊具没有特别统一的规定，都是各个厨师凭自己用什么器具顺手，就用什么炊具。但是对餐具有非常严格的规定，什么人能用什么餐具，都是按身份等级来，不能逾越。皇家用的餐具，都是由造办处出图样，在江西景德镇专门的官窑烧制，而专为皇帝和太后烧制餐具的瓷窑还各有不同。黄扒花餐具只能由皇帝使用，皇后妃子太后都不能用黄扒花餐具，因为按大清律例，只有皇帝可以用明黄色，其他任何人被发现使用明

宫廷美食器皿

黄色，都是逾越犯禁的行为，会遭到严厉的处罚。

一些专家感叹，现在的厨师对美食美器的讲究几乎一点都不懂了，即使能提供宫廷菜的餐馆也不一定提供配套的餐具。但是也有人认为，宫廷菜过于注重形式和品级，沿袭宫廷菜，应该注重其内容，食材和工序一定要严格按照宫廷菜的规矩来，但是并不需要让餐具来喧宾夺主。毕竟现在欣赏美味佳肴的客人大多数对餐具并不很在意。不过专家们仍然认为，做宫廷菜的餐馆可以保留一两个房间按宫廷风格布置，用仿制的皇家餐具，让有兴趣的客人体验地道的宫廷味。

宫廷膳房

清代宫廷的膳食管理最高机构是内务府，它下边设有两个机构，光禄寺和御茶膳房。光禄寺主管大型国家级的宴席，包括宴请外藩、宴请大臣，如重华宫小宴、蒙古外藩宴，千叟宴，等等，都由光禄寺主持和管理。光禄寺做东西也相当好，经常冠以内务府的名义，叫内府菜。皇上出席的宫廷宴席还要奏乐。御茶膳房是管皇上和他的家属吃饭的，底下有御茶房，做茶和各种茶道，再就是管做饭的御膳房，包括很多机构，分了好多局，比如做各种主食的饭局，做各种点心的饽饽局，管酒、饮料的酒醋局，等等。

紫禁城里所有的宫廷膳房都由御膳房负责，它也遵循一套严格的等级制度。御膳房就是专给皇上做饭，它做的饭菜其他人无论是皇后还是太后，都无权享用。皇宫里的其他所有有地位的人都有各自的膳房，太后的膳房叫寿膳房，妃子的膳房叫主子膳房。此外还有一个外膳房，负责临时来皇上这里开会的大臣们的膳食。

清代宫廷规定御膳房的编制是有形制的，各时期也有差别，晚清时期达到370多人，包括尚膳正、尚膳副、尚膳、庖长、副庖长、庖人、承应长、承应人、催长、领催、厨役等。对其他膳房的人员编制没有规定，但是有供货编制。比方太后的寿膳房一天供多少猪，多少羊，多少牛，多少鸡，分量就是这么些，当然都是绰绰有余的。只是到了晚清的时候，因为慈禧垂帘听政，所以寿膳房人员和御膳房人员差不太多。

御膳房的几百位御厨每个人各有一灶，专做一道大菜，叫做"一菜一灶"。给皇上做的所有的菜都是必须量化的，每种油盐酱醋原材料都量化得特别准确，做之前一定称准了，不许多，不许少，都是标准化，每天都一样，不允许有一点改变。即使皇上和太后皇后妃子一起用餐，他也是只用他自己桌上的菜，不许任何人上这个桌吃饭。除非皇上把御膳房专给他做的菜赏给别人，其他任何皇家成员都不可以动这些菜。

因为给皇上做的菜都是量化的，所以这些菜并不好吃。御膳房的菜要定量，都是要称的，保证每一天做的菜标准都一样，所以都是事先做好的，保证皇上随时一传膳就能上。每一道菜在厨房事先做好，放在保温的暖锅里，用大黄龙布盖上。什么时候"传膳"，端着就上去，一两分钟就得。不然一二百道菜，皇上现传膳现炒，根本就来不及。

由于这些菜是按照一个模式事先准备好而不是现炒的，所以多为炖菜、熬菜、蒸菜之类。用料再精，久吃也没法好吃。哪怕主子膳房做的菜也比给皇上做的菜好吃，因为每个主子一顿饭给皇上送菜也就两三道菜，可以现做。

故宫御厨房

御厨的外快

雍正年间任过大学士和内务府员外郎的鄂尔泰（1677-1745）及时任户部尚书的张廷玉（1672-1755）共同编纂的《国朝宫史》中记录着，当时皇太后每日的饮食原料包括："猪一口（盘肉用，重五十斤）、羊一只、鸡鸭各一只、新粳米二升、黄老米一升五合、高丽江米三升、粳米粉三斤、白面一斤、荞麦面一斤、麦子粉一斤、豌豆折三合，芝麻一合五勺、白糖二斤一两、盆糖八两、蜂蜜八两、核桃仁四两、松仁二钱、枸杞四两、晒干枣十两、猪肉十二斤、香油三斤十两、鸡蛋二十个、面筋一斤八两、豆腐二斤、粉锅渣一斤、甜酱二斤十二两、清酱二两、醋五两、鲜菜十五斤、茄子二十个、王瓜二十条"。皇后的差不多减半，但也很不少了。

光绪年间御膳房的厨师陈光寿先生所传其徒弟王殿臣等人的菜折子则记载着，乾隆皇帝（1711-1799）的一顿早餐可以有18道菜，包括肥鸡锅烧鸭子云片豆腐、燕窝、鹿筋狍肉、三鲜丸子、奶白粥、酱黄瓜、苏油茄子及点心等。陈老先生曾跟家人回忆说，皇上的午餐晚餐会有40道菜，通常是70道，最多达200道，都是一块儿上。这些菜会摆在好几个甚至十几个桌子上，皇上常吃的菜摆在离他最近的桌子上，一伸筷子就能够到。

这么多菜皇上一个人当然吃不完。每餐饭下来，大多数菜动都没动过。这些菜，除非皇上赏赐给其他人，就都归皇上身边的太监和御厨处理。这成了这些太监和御厨们可以挣外快的一种特权。这些菜端回御膳房之后，赶紧装车，送出紫禁城，送到皇宫附近一些和他们关系密切的上等餐馆，那里正有亲王贝勒皇亲国戚等着享受这些个菜，这些菜当然都是高价卖的。卖这些菜所得的收入由参与的太监和御厨两下平分。

这笔外快可不少。陈光寿先生作为御膳房厨师每月的工资是6两纹银。但是把皇上吃剩的菜卖给大餐馆的外快，他实际上每天得到的至少就有6两银子。那比一品官的工资还高得多。

宫廷菜在清末民初达到了一个高峰，把御膳菜卖到馆子里的生意一时也特别兴旺。这个生意在民国初年仍然兴旺，是因为虽然1911年的辛亥革命推翻了满清帝国，末代皇帝溥仪逊位，但是民国政府每年还要给溥仪一部分补贴，溥仪还留在紫禁城的后宫里自称皇上，那时候御膳房还有，他也还能保持他帝王生活的水准。但是1924年冯玉祥把溥仪撵出了皇宫，御膳房和其他的宫廷膳房就不再存在了，从而也结束了兴盛一时的卖御膳菜的生意。

会贤堂是老北京著名八大堂之一

但是这时候，宫廷菜已经走出了紫禁城的红墙，与北京高档餐馆的厨艺融合为一体。即使随着前清王子王孙的没落，靠这些人吃饭而盛极一时的饭庄也走向衰败，很快就有新的高档餐馆取代了它们，依然还是为上等人服务。

新一拨的餐馆仍然保持着很高档次的烹饪技术，这主要是仰仗它们的厨师。这些厨师很多都是御厨人家的子弟。陈光寿唯一的女婿王殿臣，就成了上世纪二三十年代北京首屈一指的餐馆致美楼的掌灶厨师。这些厨师都具有御膳房厨师的一些本事，每个人都至少会做几十道御膳菜。宫廷菜由此传到了民间。

御厨

御厨基本上都是家传。他们没有食谱，也没有任何的教科书，完全是口传心授。学徒主要靠个人的经验和领悟能力。

宫廷菜的厨师是子承父业，进宫前必须在宫外学徒出师后才能继承父业。学徒要学习选料、切配、掌握火候、操作、制作的色香味型以及菜品的装饰特色，在实际学习中还要亲自品尝口味，悟性好的很快就能上道，做出来的菜味儿就特别好。

像陈光寿老先生这样有心的御厨总会收集宫廷的菜单，记录在一个折子上。这些菜折子只记着菜名，并没有烹饪的方法步骤。陈光寿给女婿王殿臣留下了八本这样的菜折子，但是现在只剩下一本了。

有意思的是，这类传承的菜折子中有的正面是菜谱，背面却写着一个个药方子，都是治疑难杂症的偏方，也是这位御厨收集的。这是因为当时的厨师非常讲

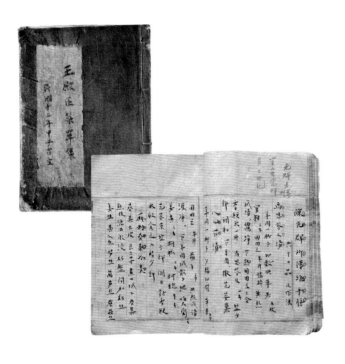

经王殿臣等人整理的陈光寿所传菜折子

究道德。他们自己不喜欢杀生，但是厨师这个行业却使他们不得不为之。因此他们心里总是有对生命的一种愧疚感。所以每个厨师都在抄药方子。杀一个生，就送一个抄的偏方给需要治病的人。

这样的职业操守同样是宫廷烹饪文化一个必不可少的组成部分。

西法菜

陈光寿老先生抄录的几百个宫廷菜都是固定的。在清朝，宫廷菜是不能发展的。所规定的配料多少，谁也不能改。给皇上吃的菜都是入了典籍的，变化只能是量多量少的变化。皇上要换口味吃新菜，可以上八大楼要菜。皇后和妃子也不能和皇上同桌用膳，除非皇上打赏，也不能吃皇上御膳房做的菜。但是她们可以做菜送给皇上。所以，宫廷菜发展不了，到清朝灭亡，宫廷都没有了，宫廷菜就更无从发展了。

皇上改换口味的一个机会就是离开北京去外地巡视的时候。不过清朝有这种机会的皇帝没有几个。其中一位是曾经六下江南的乾隆皇帝。每次出巡之前几个月，御膳房就派人去皇帝主要的落脚行宫指导培训当地的厨师做宫廷菜，准备好一百多道菜的菜单子。一旦皇上到了地方，他仍然吃的是御膳房的菜。但是他的几千随行人员得在别的地方用膳，通常是寺庙这类场所。为这些随从做饭的则以当地厨师为主，做的也是他们擅长的当地名菜。随行的官员吃哪些菜觉得比较好，就会给皇上推荐，其中一些菜连带其厨师最后都到了北京。其中典型的一道菜就是苏造肉，成了宫廷菜里少有的几个苏菜之一。

也是从乾隆开始，给皇上摆桌的时候，就要在左边摆一副刀叉了。不过这和吃西餐没有关系，只是为了吃葱烧海参这类菜肴方便。宫廷里吃西餐是在光绪皇帝年间，但是还是外送，御膳房还是不做西餐。

西法菜主要由掌灶厨师做，因此本质上还是中餐。当时北京的西餐馆叫番菜馆，无权往宫廷里送菜。所谓西法菜是西式做法，使用的原料是中西搭配。有好多原料都是用中国的东西代替。比如西芹，就用中国芹菜。这些菜做得也都很好，毕竟掌灶厨师的水平摆在那里。西餐就这样进入了北京的高档餐馆，但是还不能算作宫廷菜。

Introduction to Royal Chinese Cuisine

Xiong Lei

Beginning with the earliest emperors of ancient times, many palaces have been built throughout China's history, and naturally many royal dishes were created. But the royal dishes most commonly found today are predominantly those of the Qing Dynasty(1644-1911). As the last of China's feudal dynasties, as well as one in which an ethnic minority was in power, Qing royal dishes carried on many characteristics of those from the Ming and previous dynasties, but also introduced the dining customs of ethnic minorities such as Mongolian, Hui and Manchu, forming the ultimate combination between royal cuisine and that of the common people.

The Foundation of Royal Dishes

The cornerstone of Qing royal dishes was Lu cuisine. Originating from the peninsular Shandong Province (known as the State of Lu in ancient times), Lu cuisine is one of the eight major regional styles of Chinese cooking. The home of

Confucian manor

Confucius, who highly valued cuisine and dining etiquette, the culinary arts have a long history in Shandong. Lu cuisine reached maturity during the Yuan Dynasty(1279-1368), but its roots can be traced back more than 3,000 years to the Spring and Autumn period. Not only does Shandong border on the ocean, it also features widely varied terrain, including mainly grasslands and mountains, and produces a wide array of products, from sea food and sea salt to all types of grains and cereals, therefore historically the people of Shandong had access to many different foods and seasonings, creating a firm foundation for the development of Lu cuisine.

Due to innumerable incidents such as war and famine throughout history, many people of Shandong have been forced to leave their homes, relocating to other areas of northern China, such as Beijing, Tianjin and the northeast. Some of these migrant Shandong residents became royal chefs during the Qing Dynasty.

Qing Dynasty chefs were all of Han ethnicity. The Manchu people would not have other Manchus cook for them, as they believed this was a job for people of lower social status. After the Manchu people rose to power, all Manchus enjoyed royal life, as they received generous amounts of food and money without having to labor. Therefore it would be impossible to imagine a Manchu cooking for others.

Lu cuisine was not only prominent among Qing royal dishes. According to sources, when the third Qing emperor, Shunzhi (1368-1661), first relocated from Manchuria in northeastern China to Beijing, he brought with him to the Forbidden City a total of seven chefs from the royal kitchen of the recently overthrown Ming Dynasty (1368-1644). All seven of these royal chefs were from Shandong. This signifies that the cuisine of the Ming Dynasty, the last dynasty ruled by the Han people, predominantly featured Lu dishes.

So, with Lu cuisine as its foundation, Qing royal dishes also assimilated and enhanced by Manchu dishes that the Manchu people were accustomed to eating in the northeast. The majority of Manchu dishes used northeastern ingredients, such as venison, bear paws, quail, and so on, all of which are authentic northeastern dishes, as well as typical "old-fashioned" royal dishes.

Another representative Manchu delicacy is bobo, a type of pastry, which also forms an important part of royal cuisine. When the Manchus were in the northeast they were originally a nomadic hunting people, but they took pride in being able to make all kinds of pastries. Whenever they held large feasts, wealthy tribes would compete by covering tables with bobo, and whoever could stack their table the highest was regarded as the richest tribe. Unfortunately not many Manchu pastries can still be found today, but those that can are still greatly enjoyed, such as candied fritter, lotus cake, yogurt rolls, and so on.

Spare No Costs

Royal dishes are based on Lu cuisine, but more refined, and could be said to "spare no costs". If the best Beijing zhuangguan cuisine restaurant used a total of two pounds of ingredients to cook one pound of meat, then royal cuisine would use eight pounds. And all ingredients must be the best and most expensive, paying no heed to ingredients of "substandard quality". For example, when cooking the dish "longevity braised pork" for the emperor, the meat was required to contain eight layers of both lean and fatty meat, and the lean meat was required to be twice as thick as the fatty meat. In order to obtain a piece of meat suitable for this dish, at least a dozen pigs had to be slaughtered, and several layers of the alternating lean and fatty meat had to be "made by hand".

Another extravagant dish was "yellow croaker stuffed with hot gravy". The fish itself cost more than 1,000 yuan, and the fish's belly was stuffed with some of the most expensive foods ever consumed by humans, such as edible bird's nest, shark fin, abalone and sea cucumber. The soup in which the fish was cooked was equally demanding. And the cooking process of the dish was extremely cumbersome, as the fish had to be steamed, then boiled, then fried; if the shape of the fish was compromised in any way then it would have to be discarded and replaced with a new one. In total just to cook a single fish it would cost a total of about 5,000 yuan. That's why today most restaurants cannot afford to make royal dishes, and for ones like JE Mansion that still cook such dishes, there is not much money to be earned from dishes like yellow croaker stuffed with hot gravy, due to the lavish cost of its ingredients and arduous preparation process.

Another important element is the dining ware and cutlery, which experts state is essential to royal cuisine culture, as "beautiful dishes must be accompanied by beautiful dining ware". In ancient times, it was not uncommon to design and produce a certain type of plate or bowl for a particular dish.

In terms of cooking tools there were no rules or traditions in Qing Dynasty royal cuisine, and each chef just used what he was most comfortable with. But when dining there was an extremely complex set of regulations, including the type of dining ware used by each person, which was based on his or her social status. For those used in the imperial palace, the porcelain was shipped in from Jingdezhen of Jiangxi Province, and each plate made for the emperor and empress was required to be unique in design. Dining ware which was yellow in color was reserved for the emperor, as yellow was the emperor's color ("yellow" being a homonym of "emperor"), and not even the empress, royal concubines or princes could use anything yellow; according to Qing law, if this rule was broken by anyone, there would be dire consequences.

Some experts have pointed out that most modern chefs are completely oblivious to the etiquette involved in matching dining ware with dishes, and even most so-called royal cuisine restaurants do not provide properly matching porcelain. But some people believe that overemphasizing the form leads to underemphasizing the content, and that royal dishes should be made strictly according to traditional methods, with all the proper ingredients, but it is not necessary to attract diners with intricate porcelain designs. After all, most customers are there to enjoy the food, and are not aware of or concerned with the dining ware used. However, experts believe that royal cuisine restaurants should still have one or two rooms furnished according to the royal palace design, complete with historically accurate dining ware, to allow customers who are interested to enjoy a completely royal experience.

Royal dining ware

The Royal Kitchen

In the food management system of the royal palace the highest authority was the Office of Internal Affairs, followed by the Guanglu Temple and Royal Tea and Cuisine Kitchen. The Guanglu Temple (actually the name of a team of chefs) was in charge of large-scale national-level banquets, such as those held for representatives or ministers of foreign countries. The team was capable of preparing extremely elaborate and delicious meals, often customized for guests from a certain region (such as Mongolia), and would often serve meals under the name of the Office of Internal Affairs, thus this type of cuisine became known as Internal Cuisine. Any banquets where the emperor was present were required to be accompanied by music. The Royal Tea and Cuisine Kitchen were responsible for cooking for the emperor and his family, as well as preparing royal tea ceremonies. Each of these teams was further divided into several sectors, each responsible for staple foods, pastries, beverages, and so on.

In the Forbidden City all palace kitchens were under control of the main royal kitchen, and were required to strictly follow a set of status differentiation regulations. The Royal Cuisine Kitchen cooked only for the emperor himself, and no one was allowed to have a bite of any dishes made there, not even the empress or prince. Each additional social rank had its own kitchen, such as those reserved for cooking for the empress, royal concubines and visiting ministers.

The Qing Dynasty royal palace had structured regulations for cooking and dining, and each period differed as well. In the late Qing the royal kitchen involved more than 370 people, including cooks, cook assistants, head chefs,

Forbidden City royal kitchen

second head chefs, as well as people whose job it was to tell others to rush certain dishes, someone else in charge of the dish rushers, and so on. There was no rule for how many people were involved in the cooking process, but there were requirements for amounts of food, such as in the empress' kitchen how many pigs, sheep, cows and chickens were allowed per day, the daily limit for which of course was appropriately excessive. By the late Qing Dynasty, when Empress Dowager Cixi was in effective power, the number of staff employed in the empress' kitchen was more or less as high as that in the emperor's.

Each of the several hundred chefs in the royal kitchen had his own stove to cook a particular dish on. The ingredients for all dishes served to the emperor were required to be accurately measured before the dish was made, and the standard measurements had to be strictly followed every day. Even when the emperor ate with the empress and royal concubines, he only ate from the dishes on his own table, and no one was able to try a single bite of the emperor's dish, with the exception that he had his kitchen make a dish for a particular person.

Since all the emperor's dishes had standardized amounts of ingredients such as oil and salt, they didn't necessarily taste as good as possible. All the emperor's dishes were made in advance, and held in a warm pot, to ensure that the emperor could eat at any time. Whenever the emperor announced he wanted to eat, a yellow dragon tablecloth would be spread, and his dishes would be ready in a minute or two. Otherwise, to make more than a hundred dishes on the spot from scratch would take a long time, and that simply would not do.

Since the emperor's dishes were all prepared in advance and waiting at the ready, most were stewed, boiled or steamed. Even with the finest ingredients, many dishes could not be preserved for too long. As a result, dishes made in other kitchens actually tasted better than the emperor's, such as those made for the concubines, who were only entitled to two or three dishes per meal, as they could be made upon request.

Wang Dianchen and Mr. Chen's other disciples arranged the recipe notes that he passed on to them

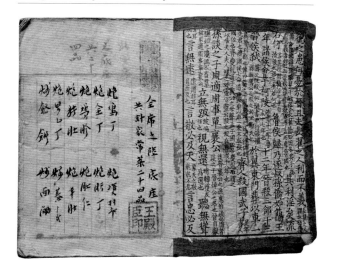

The Emperor's Leftovers

According to records in Feudal Palace History, compiled in cooperation by Ertai (1677-1745), who during the Qing Dynasty Yongzheng era (1723-1736) had acted as Grand Secretary and Minister of Internal Affairs, and Zhang Tingyu (1672-1755), who at the time served as Minister of Revenue, in their time the emperor's daily meals consisted of the following materials and ingredients: "1 pig (for meat served on a platter, 50 jin); 1 sheep; 1 chicken; 1 duck; 2 sheng of polished round-grained rice; 1 sheng 5 ge of aged yellow rice; 3 sheng of gaolijiang rice; 3 jin of polished round-grained rice flour; 5 jin of white flour; 1 jin of buckwheat flour; 1 jin of whole wheat flour; 3 ge of peas; 1 ge 4 spoons of sesame seeds; 2 jin 1 liang of white sugar; 2 liang of dish sugar; 8 liang of honey; 4 liang of walnuts; 2 qian of pine nuts; 4 liang of wolfberries; 10 liang of dried dates; 20 jin of pork; 3 jin 10 liang of sesame oil; 20 chicken eggs; 1 jin 8 liang of wheat gluten; 2 jin of tofu; 1 jin of fenguozha; 2 jin 12 liang of sweet sauce; 2 liang of mild sauce; 5 liang of vinegar; 15 jin of fresh vegetables; 20 eggplants; 20 cucumbers." (1 jin, or catty, is equivalent to 500 g; 1 sheng is approximately equivalent to 1 L; 1 ge is equivalent to 0.1 L. 1 liang is equivalent to 31.25 g; 1 qian is equivalent to 3.125 g.) The empress' meal consisted of about half this, which is still much more than one person could ever possibly eat in one meal.

According to the notes Mr. Chen Guangshou passed on to Wang Dianchen and his other disciples, taken when he was working as a chef in the royal kitchen during the Guangxu period (1875-1908), the breakfast of the Qianlong Emperor (1711-1799) consisted of 18 dishes, including fatty chicken and braised duck cloud cut tofu, edible bird's nest, deer tendon and roe deer meat, three flavor meatballs, white congee with cream, cucumbers with soybean paste, crispy eggplant, pastries, and so on. Mr. Chen once recalled to his family members that the emperor's lunch would consist of at least 40 dishes, often 70, at the most 200, all of which were served at the same time. These dishes would be arranged on several or even more than a dozen tables, and the emperor would eat from the dishes that were on the table closest to him, and within chopstick grasping range.

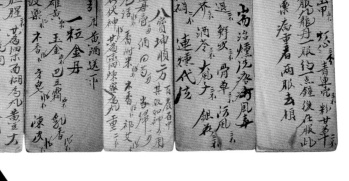

Royal recipes collected by Mr. Chen Guangshou

Of course the emperor could not come anywhere close to finishing all of these dishes. After each meal, the majority of the dishes would not have even been touched. Unless the emperor chose to offer any of these dishes to someone in particular, it would be up to the eunuchs and chefs to take care of them. These imperial leftovers became a source of extra income for them: as soon as the dishes were sent back to the kitchen, they would be shipped off on a cart out of the Forbidden City, to some of the nearby high-end restaurants with whom the eunuchs and chefs were acquainted. At these establishments slightly more distant relatives of the emperor could enjoy royal dishes, for a high price of course. The profit made would be split between the eunuchs and chefs.

This extra bonus was quite a large sum of money. As a chef of the emperor's kitchen, Mr. Chen's monthly salary was six liang of silver. But the money he made from selling the emperor's leftovers was six liang per day. This was much more than the salary of even the highest ranking officials.

Royal cuisine reached its peak of development at the end of the Qing Dynasty and beginning of the Republican period, when the profits of selling imperial dishes to restaurants reached an all-time high. These business was still booming in the early Republican days, because even after the Manchu Qing throne had been overtaken as a result of the Rebellion of 1911, and Emperor Puyi had been abdicated, the Republican Government still had to provide Puyi with an annual subsidy, and Puyi continued to live in the Forbidden City, where he called himself emperor, and since the royal kitchen was still in use, Puyi could still eat like an emperor. But in 1924 when Feng Yuxiang forced Puyi out of the palace, the royal kitchen and other palace kitchens fell out of use, thus forever ending the profitable business of selling dishes untouched by the emperor.

But by this time, these royal recipes had already made their way out of the palace walls, and had been combined with the previously existing menus of Beijing's high-scale restaurants. After the Qing royal family's descendants

Hui Xian Tang was one of the "Eight Great Restaurants" of Beijing

gradually fell into obscurity, the restaurants relying on these royal dishes began to disappear, and were replaced by a new generation of high-class restaurants to serve wealthy customers.

The new wave of restaurants continued to practice advanced cooking techniques, which was mainly the responsibility of the chefs. These chefs were descendants of those from the royal kitchens. Mr. Chen's only son-in-law, Wang Dianchen, became the head chef of Beijing's top restaurant in the 1920s and 1930s, Zhimeilou. These chefs all possessed the skills of the royal chefs, and each could make at least several dozen royal dishes. This is how royal cuisine was passed on to the common people.

The Royal Chefs

The recipes for royal dishes were primarily handed down from one generation to the next without cookbooks or manuals, completely by word of mouth. Disciples relied mainly on their personal experience and comprehension ability.

The recipes of royal dishes were only passed on directly from master to disciple, and only those who had previously learned from a teacher outside the palace were allowed to become the disciple of a royal chef. The disciples were required to learn how to select and prepare ingredients, how to control heat, and how to prepare dishes and their presentation. While learning the students also had to taste their own dishes; talented disciples could learn quite quickly, and make delicious dishes.

A careful-minded royal chef like Mr. Chen would collect palace menus, and record them in small accordion-style notebooks. The strips of paper contained only the name of the dish, without any details of how to make it, to prevent the secret recipe from being known by others. Mr. Chen gave his son-in-law eight of these booklets, but today only one remains.

What's interesting about these slips is that on the front of them are recipes, but on the back are prescriptions for all kinds of illnesses. This is because the chefs at the time had very high morals, and were actually quite opposed to killing animals for food, but because of their career choice they had no other option. In a way they always had a feeling of guilt towards life, so all chefs would jot down any prescriptions they came across. This way if someone they knew of was sick, maybe they could save a life to make up for the ones they had taken.

This work ethic is also an essential component of royal cuisine culture.

The several hundred dishes which Mr. Chen recorded were all standardized, as during the Qing Dynasty dishes were not allowed to change or develop. The regulated amounts of ingredients could not be changed by anyone. Dishes served to the emperor had been added to imperial records, thus only the amount of the dish could vary, not the proportions. If the emperor wanted to try something new, could visit one of the "eight major restaurants" of Beijing. There he and his empress and concubines could eat at the same table, and although none of them could eat the emperor's dishes (unless offered to them), the empress and concubines could order dishes for the emperor to try. So royal dishes were not able to develop to begin with, and after the fall of the Qing Dynasty, even the palace life had disappeared, thus royal dishes had even less hope of evolving.

One chance the emperor had to try new dishes was when he left Beijing and traveled to other regions of his territory. But in the Qing Dynasty very few emperors had this opportunity, among which the most famous is Emperor Qianlong, who visited southern China on a total of six occasions. Several months before each tour, the royal kitchen would send people to the main stops along the path the emperor would be traveling, to teach the local chefs how to make royal dishes, of which they would have to prepare at least 100 types of dishes. So wherever the emperor visited, he would still eat the same meals as he did back at the palace.

But the several thousand people in his caravan would have to eat elsewhere, such as at temples. The food eaten by the entourage would merely consist of what the local chefs normally cooked. If the members of the entourage thought any dish was particularly good, they could recommend it to the emperor, and if the emperor liked it he would take the cook and any necessary local materials with him back to the palace. A typical example of this is Suzao pork, which is one of the very few Suzhou dishes on the royal menu.

Beginning from Qianlong's time, when setting the table for the emperor, a knife and fork would be included as well. But this was totally unrelated to western cuisine, it was simply for the convenience of eating foods such as onion braised sea cucumber. It wasn't until the Guangxu Emperor's time that western food began being served in the palace, but all western dishes were sent from outside, as the royal kitchen never made western food.

In Beijing at the time western dishes could be eaten at high-end Chinese restaurants, where they were known as "western-style dishes", but they were still made by the same chefs, so they were still essentially Chinese dishes. Western restaurants were called "foreign restaurants", which were not authorized to send dishes to the palace. So the "western" dishes which the emperor was able to eat were made with western cooking methods but a blend of Chinese and western ingredients, and many main ingredients were replaced with the nearest equivalent in Chinese cuisine. For example Chinese celery was used instead of the thicker-stalked celery commonly found in the west. Despite the lack of authenticity, these dishes were still quite delicious, as the head chef was very highly skilled. This is how western food appeared in the high-class restaurants of Qing Dynasty Beijing, but never truly entered the royal palace.

健壹集团及其公馆菜系渊源

健壹集团以其创办人康健一的名字命名，取自"自强不息天行健，厚德留芳心归一"之意。

健壹集团是一家综合性企业，旗下公司涉及餐饮酒店、影视投资、房地产开发、贸易等行业，包括北京健壹公馆、健壹景园、大宅门迎祥商务酒店、上海科学会堂1号楼JE及四川大宅门酒业有限公司。

健壹集团董事长康健一先生一直主张，作为一名中国人和中国企业家，他本人和健壹集团有责任和义务为中国文化的传承和向世界的推广作出贡献。健壹集团旗下所有企业都有着极为鲜明的文化特色，并把文化保持和文化融合始终贯穿于经营管理之中。

以集历代御膳菜之大成的清朝宫廷菜和清末民初北京高档庄馆菜为基础的健壹集团公馆菜系，就是健壹集团推介中国文化传统的一种努力，也是健壹集团餐饮业的成功之道。健壹公馆曾于2008年被世界酒店联盟授予中国最具特色精品酒店品牌，并成为中国达到加入世界精品酒店行业最高端的欧洲城堡与休闲联盟餐饮标准的仅有的三家酒店当中的一家。

公馆由来

公馆在中国自古就有。古代公馆特指公家所建造的馆舍，如诸侯的离宫别馆或宫室。近代以来专指大官或富家的高档住所。作为达官贵人的宅邸，近代公馆兴起于清末民初。随着末代皇帝的逊位，王宫贵族没落了，昔日王府也更换了主人，新贵们的宅邸多成为公馆。很多公馆规模之大，可以容纳众多工作人员，成为一个完善的办公机关。像西安事变发生后蒋介石住过的高桂滋公馆，抗战时期中共中央南方局设在陪都重庆的办公地点周公馆等，都是有名的公馆。

近十多年来，中国一些企业家、建筑家开始打造公馆式的高档会所。康健一先生便是其中的一位。他热爱中国传统文化，学过建筑，画过国画，搞过拍卖，2001年投入到酒店餐饮业，梦想造就传世的中国风格精品酒店。他起手推出的大宅门迎祥商务酒店，就以其古色古香的建筑环境和京味儿十足的菜品广受好评。接着，他又在距北京首都国际机场25公里处的东四环路红领巾公园内，精心建造了占地5.6万平方米的健壹公馆，其后又在上海、内蒙古等地修建了同样高档次的健壹公馆。

健壹公馆及健壹集团麾下的其他酒店，以中国古典的园林建筑群为特色，有悬山顶的殿堂，有廊桥，有榭舫式的单间餐厅和客房，溪流绕屋而过，修竹森森，假山嶙峋，草坪空阔。置身其中，从落地窗向外望去，景色怡人；室内古典的中式家具和陈列的古董，无不体现出一种高雅和尊严。这样的环境，无疑会给客人们留下深刻的第一印象。

公馆菜与宫廷菜

要做"带有中国人品牌烙印的酒店"，能够集中体现中国文化的一个重要方面，便是它的菜品。

公馆菜并未被列入代表中国烹饪艺术的八大菜系之中，因为不同地方的公馆，自会融合不同地域的菜系风格。但是在公馆兴盛的20世纪二三十年代，很多公馆既

健壹景园宴会区外景

兼容南北各地的菜系风格，也因公馆主人的喜好发展起自身的特色。像六朝古都的北京，既有皇家享用的宫廷菜，也有承袭鲁菜和宫廷菜精华的高档酒楼庄馆菜，还有官府世家的宅门菜，如久负盛名的谭家菜等等。

在大宅门酒店成功的基础上，康健一希望更有所提升。他把目光投向了只有古都北京才能传承下来的宫廷菜和与之密切相关的庄馆菜。为此，健壹集团聘请了清宫御厨的传人王希富老人做顾问，专门培训厨师。第一道菜"干炸小丸子"就做了三个月才算出师。而这样高水平的顾问，也使健壹集团推出以宫廷菜为底蕴的独具特色的高档公馆菜，更有了底气。

咯炸盒

健壹集团的公馆菜

健壹集团的公馆菜系是以宫廷菜为基础，以老北京八大堂、八大楼的高档庄馆菜为范本，以豪华宅门、府第菜为补充，以优秀京菜为特色的综合高档菜系。

秉承宫廷菜的传统，健壹集团的公馆菜系在原料上就很讲究。除了原料产地、质地、大小、部位都有严格的要求之外，还注重分量的足斤足两。

健壹集团的公馆菜还恢复了一部分曾经在宫廷、宅门或京城红极一时的吃食。如京城春节习俗所用的素三供，基本早已名存实亡的炸三角、咯炸盒、素丸子；北京名菜烧蝶菜中的炸鹿尾、炸佛手、炸丸子；京城御膳名点玫瑰饼；乾隆爱吃的羊眼包子；慈禧爱吃的鹿肉酥；早年京城宴会的冷荤菜品马莲肉、杏干肉、蟹肉卷等。

注重做传统文化的健壹集团，还挖掘并推出了一批早已失传的优秀名菜。如酸菜鱼翅，是综合多种满、汉菜烹饪技术的高档鱼翅菜品，不但使用了多种精良原料，而且一道菜要有20多个工序才能完成。曾经用于祈雨祭天的茄子龙、蛤蟆鲍鱼盒，不但色、香、味独特，而且造型别致，营养丰富。清代晚期慈禧最喜爱的"攒丝锅烧鸡"，不但具有原汁原味的满菜锅烧鸡的色香味形，还按照慈禧的要求为"鸡"做出了可口的"鸡窝"、"鸡草花"和"鸡内脏"。把干炸、锅烧、凉拌、滑熘和麻、辣、苦、甜、咸等"人生五味"容于一道菜中。其他如芙蓉飞龙、万字扣肉、玉珠炸段宵、九丝羹等，也是实际失传已久或已经名存实亡的极费手的工夫菜。

根据宫廷菜和庄馆菜对"美食美器"的讲究，健壹公馆的宫廷菜不仅对菜肴的造型十分讲究，所使用的餐具虽不是仿宫廷官窑烧制，却也都色形华贵、造型古雅。有金、银、玉石、水晶、玛瑙、珊瑚、玳瑁所制，也有大量特制的精美瓷器。图案造型要求做到像盆景一样美观悦目。在造型手段上主要动用的是"围、配、镶、酿"等方法。

椰盅汤圆杏仁露

围。只有十分注重配合使用,才能达到宫廷菜在造型上与众不同的特殊要求。

以宫廷菜为基础,在用料和工序上严格按照最高标准一丝不苟,但是却并非食古不化。康健一认为,宫廷菜原汁原味,制作工艺不能改变,但呈现方式要改变,不应该让形式大于用途。所以健壹集团的公馆菜装盘不用粉彩、青花,而都是用专门烧制的高档白色瓷盘。

康健一和健壹集团将正宗宫廷菜品传承下来,是希望世人看到中华饮食文化的精髓,从而感受到华夏社会发展所达到的文明程度。这应该是比宫廷菜、公馆菜更加富有魅力的所在。

"围"就是以素围荤,以小围大,并注重利用荤素菜肴本身在色彩、质地、口味、营养成分等方面的不同特点,来协调整盘菜肴的色泽调味、烘托主料,突出主味,使两者在口味、营养等方面起到互相补益调剂的作用。

"配",就是要求将两种造型不同的原料成双结对地搭配在一起,从而赋予一种特定的寓意,如以虾球和肉馅蛋饺相配制成的"黄葵伴雪梅"。

"镶",就是指在一种原料中点缀上另一种经特殊加工的原料,使菜肴更富有逼真的形象,如"金鱼戏莲"就是用青椒、豌豆、虾尾等作为镶嵌料,再在整虾和茸泥制成的鱼形胚上镶嵌点缀而成的。

"酿",就是将精加工过的各种原料,如茸泥、丝、粒等填抹在整形原料内,使菜肴的外形更加完整饱满,滋味更加醇郁鲜美。

"围、配、镶、酿"等各种方法往往是用于同一道菜的烹制加工过程中,所以它们又常常是互相包容,兼而有之。如围中有配,配中有镶,镶中有酿,酿中有

冬菜鹿肉酥盒子

The Origins of JE Group and Its Cuisine

The name of JE Group is based on that of the founder, and is a reference to two Chinese sayings: "One must constantly strive for perfection, just as the heavens maintain vigor through movement" and "those of great virtue are all of one heart, and will live on forever."

JE Group is a comprehensive enterprise, overseeing companies in industries such as hotel and catering, film and television investment, real estate development, commerce, and so on, including Beijing JE Mansion, Dazhaimen Yingxiang Business Hotel, JE Park, Shanghai Science Association Club, and Sichuan Dazhaimen Spirits Co., Ltd.

JE Group chairman Kang Jianyi, as a Chinese citizen and entrepreneur, believes that he and JE Group have the responsibility and duty to help carry on and promote Chinese culture throughout the world. All companies in JE Group possess very distinct cultural characteristics, and merge culture throughout their managerial systems.

JE Group's "mansion cuisine", which is based mostly on royal palace cuisine and Beijing's local zhuangguan cuisine

from the late Qing Dynasty and early Republican era, is one of the ways in which JE Group is striving to promote traditional Chinese culture, and is also JE Group's path of success in the catering industry. In 2008, JE Group's cuisine was selected as among the best in all of China by the World Hotel Association, and JE Mansion became one of only three Chinese establishments to be included among the finest in the world.

Origin of JE Mansion

The concept of a "mansion" has existed in China since ancient times. These mansions are locations where imperial vassals could remain temporarily when attending to affairs away from the palace. Later the term came to refer to luxurious living quarters constructed for high-ranking officials or wealthy families. As a residence for high officials and lords, contemporary mansions began to reach prominence during the late Qing Dynasty and early Republican era. After the abdication of the final emperor, the imperial royal family fell out of importance, the mansions of monarchs underwent a replacement of ownership, and many new mansions were built. Some of these mansions were quite large and impressive, employing scores of staff, with a structured office system. Among these mansions, some of the more well known ones include the Gaoguizi Mansion which Jiang Jieshi (Chiang Kai-shek) stayed at following the Xi'an Incident of 1936, and the Zhou Mansion which was constructed as an office center for the temporary capital in Chongqing during the Sino-Japanese War.

In the past several decades, some Chinese entrepreneurs and architects have designed upscale mansion-style leisure clubs. Among these people is Mr. Kang Jianyi. He has a strong affection for traditional Chinese culture, has studied architecture, has made Chinese paintings, has been an auctioner, and since 2001 has been involved in the hotel and catering industry, with the goal of creating a fine hotel which is capable of representing Chinese culture. He began by establishing the Dazhaimen Yingxiang Business Hotel, the rustic environment and authentic Beijing cuisine of which were very well received. Next he chose Honglingjin Park on East Fourth Ring Road, 25 km from the Beijing Capital International Airport, as the location to construct JE Mansion, with a total land area of 56,000 sq m. Since this first installment, additional JE Mansions have also been constructed elsewhere in Beijing, as well as in Shanghai and Inner Mongolia.

JE Mansion, like other hotels of JE Group, features classical Chinese garden-style architecture, including palace halls on mountain tops, corridor bridges, waterside guest rooms and dining areas, streams, bamboo thickets, rock gardens, and some of the largest lawns in China's metropolitan areas. Much of this scenery can be enjoyed first-hand, or from the view in the guest rooms, and each of the rooms is decorated with a completely different selection of antiques. JE Mansion is not just a hotel, it's an experience.

Mansion Cuisine and Royal Cuisine

Creating "a hotel which is truly Chinese" would not be complete without consideration to an important aspect of Chinese culture: cuisine.

Mansion cuisine did not become listed as one of the eight main types of Chinese cuisine, as these cooking styles are based on region, and mansion cuisine assimilates influences from several regions throughout China. But from the 1920s through to the 40s, when mansions in China began to reach prominence, many mansions combined dishes and elements from throughout the north and south, and each mansion developed its own unique characteristics depending on the preferences of the owner. In Beijing for example, which was the capital for six dynasties, mansion cuisine includes royal dishes once enjoyed by the emperor himself, as well as zhuangguan cuisine which continues the traditions of late feudal luxury restaurants, as well as cuisine enjoyed by the families of government officials, such as Tan family cuisine.

After the success of Dazhaimen, Kang Jianyi wanted to take his business to the next level. He focused his attention on royal cuisine, which can only be found in Beijing, and the closely related zhuangguan cuisine. JE Group hired Qing Dynasty royal cuisine successor Wang Xifu as their adviser and culinary trainer. The chefs practiced the first dish, dry-fried meatballs, for three months before Mr. Wang approved it. Only with an adviser of such a level of expertise is JE Mansion confident in claiming they specialize in authentic royal cuisine.

JE Group's Mansion Cuisine

JE Group's mansion cuisine menu is based upon royal cuisine, according to the zhuangguan cuisine enjoyed at the "eight major restaurants" of Beijing in the late Qing Dynasty, and further complemented with dishes prepared at various mansions and official manors, forming a comprehensive representation of traditional Beijing cuisine.

In accordance with the traditions of royal cuisine, JE Mansion pays particularly careful attention to the ingredients they use. In addition to having strict demands toward production location and quality, they also ensure that the dishes are well-sized and satisfying.

JE Mansion has revitalized a series of dishes which were once enjoyed every day in Beijing, either in the royal palace, the homes of officials, or luxury restaurants spread throughout the city. For example, the once-extinct "three vegetarian dishes" (deep-fried triangles, gazha crisps and deep-fried vegetable balls) traditionally enjoyed by Beijingers during the Spring Festival; rose petal cakes, a royal kitchen specialty; sheep eye steamed buns, one of the Qianlong Emperor's favorites; Empress Dowager Cixi's beloved baked venison puffs; and various cold meat dishes once served at banquets in the ancient capital, such as braised pork with iris petals, dried apricot meat, crab rolls, and so on.

JE Mansion, with their ambition to promote traditional culture, has also researched and resurrected a series of dishes which had once been lost for ages. Among these are shark fin in pickled cabbage, which is a luxury dish combining a variety of Manchu and Han culinary skills, and involving both a large number of fine ingredients and a preparation process including more than 20 steps. There are also the eggplant dragon and frog and abalone cakes, once used in ceremonial rituals (such as praying for rain), which not only feature unique color, scent and flavor, but also have very distinguished designs, and are highly nutritious. Also on the menu is another favorite of Empress Dowager Cixi, crispy roast chicken slivers, which incorporate the original taste and form of Manchu roast chicken with special adjustments made specifically for Cixi, for a dish that combines the methods of deep-frying, quick-frying, roasting and dressing with sauce, with the flavors of spicy, bitter, sweet and salty, into a single dish. Other rare dishes that have disappeared from the table, and some even from historical records, include lotus flower dragons, longevity pork, jade bead clouds, royal congee, and many others.

According to the "fine cuisine, fine dining ware" etiquette of royal and zhuangguan dishes, not only is JE Mansion very particular about the dishes themselves, but they also use very exquisite classical-style dining ware to serve them. Among the materials used are gold, silver, jade, crystal, amber, coral and hawksbill, and large numbers of fine

Lobster with rice steeped in sour soup

porcelain dishes were also custom-created as well. The create dining ware is just as much a work of art as the meal. In order to create dishes which are very aesthetically pleasing, a set of special skills are used, namely "surrounding", "pairing", "decorating" and "rising".

"Surrounding" refers to surrounding meat with vegetarian ingredients, or larger portions with smaller ones, as well as utilizing the unique characteristics of aspects such as the original colors, textures, flavors and nutritional elements of foods, to adjust the overall appearance and flavor of the dish, thereby highlighting the main ingredients and taste, and creating the most combination of all its aspects.

"Pairing" refers to combining two different ingredients to form a specified meaning, such as shrimp balls and meat and egg dumplings, which together become "musk mallow and snow plum".

"Decorating" is adding to a dish an ingredient that has undergone specialized processing, to make a dish become more visually representative of a concept. An example includes "the goldfish plays among the lotus flowers", which uses green peppers, peas and shrimp tails as the decorative ingredients, which are then placed atop full shrimps and meat paste arranged in the shape of a goldfish.

"Rising" involves adding additional components, in forms such as paste, shreds or grains, to main ingredients to increase the fullness of a dish's appearance and the completeness of its flavor.

These four special skills are often used in combination when preparing a dish to increase its overall qualities and appearance. Only through strict adherence to the etiquette involved in these methods will one be able to achieve the proper design of traditional royal cuisine.

JE Group's cuisine is based on royal cuisine, with ingredients and preparation strictly adhering to the highest level of standard, but the manner in which the dishes are served is different from ancient times. Kang Jianyi believes that the taste and culinary practices of royal cuisine must remain the same, but the fancy dishes used in ancient times are no longer necessary. To retain emphasis on the food itself, all mansion cuisine prepared by the JE Group is served on custom-made plain white porcelain dining ware.

Kang Jianyi and JE Group offer these royal dishes to visitors from around the world, and hope that from them others will be able to experience the rich history and culture of China, which Kang believes is much more meaningful than the traditions of mansion or royal cuisine alone.

把一件事做到极致

走近康健一

每年大年初一，他都会到北京故宫走走看看。缓步于紫禁城，他会心生将太和殿复制到健壹公馆草坪上的念头。这是健壹集团董事长康健一的过年习俗。

作为土生土长的北京人，康健一对传统文化、艺术和历史，可以说爱到了骨子里。就连他的名字，也透着浓浓的中国传统文化的韵味。父母给他取名康健，已经是很不错的意境，象征着身心的健康。但是2003年，他给自己的名字后面加上了一个"一"字，既体现了中国古代哲人老子的道及万物生于"一"的哲理，也表明了他"自强不息天行健，厚德留芳心归一"的心愿。

在学过国画，做过拍卖，搞过收藏之后，康健一于2001年开始投身酒店餐饮业。开始只是出于"喜欢"，也是"机缘巧合"。他说，"夫人家是做餐饮，机缘巧合就带到这个路口了。"即使看似随意而为，但是融到骨子里的那种对传统文化的痴迷，还是让康健一不由自主地要让他的酒店体现出一种与众不同的文化品位。于是在大宅门迎祥商务酒店之后，又有了中国传统文化特色更加浓郁的健壹公馆。

从现代酒店业引进中国的一个世纪以来，特别是上世纪70年代末中国开始有了合资酒店以来，高档豪华饭店已经如雨后春笋般遍及华夏大地。可是康健一却发现了它们的不足。"就像我到法国希望看的是古堡、卢浮宫一样，外国人到中国以后希望看到的也是中国传统的一面。"康健一说。可是中国众多的星级酒店，哪怕是园林式的建筑，传递出中国传统文化信息的也并不多见。毕竟绝大多数酒店只是以盈利为目的，赚钱就好。

而醉心于历史和文化的康健一，却把酒店作为文化事业来做。即使成功地做过拍卖，又成功地推出了好几处以自己名字命名的高档公馆，这位健壹集团的董事长也从不

健壹公馆外景

以商人自居。他说自己不是在做生意,而是在做艺术。

因为是做艺术,所以康健一做酒店,做公馆,"纯粹就是要做自己喜欢的事",而很少考量投入与回报的比例。得益于多年做酒店的管理经验,他说从开始筹备到现在"没有遇到什么大的麻烦和困难,一切都很顺利"。只是有一点他坚定不移:把古今、中西的文化及艺术与现实生活完美结合,做一个让中国人骄傲,让外国人尊重的酒店。

这是康健一的梦想。但是康健一没有耽于梦想,而是将梦想付诸行动。"健壹公馆表达了我对于艺术的感悟以及对于中国传统文化的理解,希望通过它促进东西方文化间的高品质交流。"康健一说。

当人们走进健壹公馆,感觉就好像走进了一座博物馆。大堂陈列着清朝皇帝的龙袍,古朴的砖雕石雕镶嵌在低调却不失典雅的石板墙或地板之间,明清家具和欧式沙发相得益彰地摆放在同一个空间里。它们很多都是康健一多年的收藏。更独特的是公馆各个房间的上百个门扇,几乎全部来自他从全国各地的民间收藏。这些门扇,贵的上万元,便宜的也要两千多元。"每扇门都有自己的故事,都有着自己的年代和人家。"康健一说,"别人的艺术品是用来观赏的,我这个艺术品是可以享用的。现在北京有很多复古餐厅,他们的老门扇都只是一种装饰,而我这里每个房间安装的老门扇,全部都是真正的使用。"这正是"人弃我取、人取我予"的精神。

博物馆的感觉和这些收藏品的使用,正是康健一把文化、艺术和现实生活完美结合起来的一种实践。"好东西放在库房里没有任何意义，惟有分享才能体现出其最大价值。"他说，"博物馆的东西只可'远观'，不可'亵玩'，咱家的东西既能看又能用。"

健壹公馆，正是可以"享用"的博物馆。

独一无二

一个成功的企业和经营模式，不被模仿抄袭是不可能的。对此康健一很有自信——别人可以拿了尺子来量，他说，"我们不怕被模仿。因为一个成功的企业一定是有灵魂的，那些精髓的东西是无法被复制的，即使做到相似，很多细节的地方也还是照搬不走的。"

他的自信，源自他的一个理念：作为服务企业，其精髓不仅在于建筑，更重要的是服务理念和态度，是给客人独一无二的感受。

康健一追求把事情做到极致，就是要做到独一无二，或者说绝对的唯一。健壹集团的酒店就体现了这种追求。绝对中式的环境不够舒适，绝对西式的东西又没有中国的文化。所以，健壹公馆是中华文化和西方的生活方式的结合，从建筑到室内陈列到服务，处处体现了中国的文化，却又和现代家居的舒适完美地融为一体。

最初他也认为一个酒店最能吸引人的应该是其独特的建筑。随着健壹公馆的发展和自己眼界的开阔，康健一逐渐意识到，健壹公馆和世界各地的其他建筑是没有可比性的，因为各地的文化造成了各地都有其独特的建筑，是不能放在一起作比较分出谁好谁坏的。

"做中式风格的精品酒店，并不是将含有中国元素的物件简单堆砌，更重要的是在点滴间释放中国文化博大精深的内涵。"康健一说。健壹公馆中艺术珍品或悬挂或摆放，像是漫不经心，却又给人一种家的感觉。"公馆其实就是家的意思，并且主人把认为美的东西和大家一起分享。"康健一说，"希望来到健壹公馆的人都能拿这里当家，感受到家的温暖。"说到底，"酒店是给人住的，而不是给人看的。"

出于这样的认知，康健一虽然一直延续着做中式文化的概念，但是越做越简约了。"现在学会减法，就越做越轻松，不像原来就是一味地堆砌，满眼都是文化，没有重点。"他说。

经过几年的磨练，健壹公馆做出了自己的特色。第一便是餐饮。健壹公馆以宫廷菜为基础的菜系水准非常高，已经加入了RELAIS & CHATEAUX（罗莱夏朵精品酒店集团），是世界精品酒店行业联盟里一个最高端的组织。目前中国能达到这个组织餐饮标准的酒店只有三家，健壹公馆就是其中之一。第二就是环境及建筑。健壹公馆的环境是不可复制的，在城市中闹中取静，其珍贵性不言而喻。第三就是服务。健壹公馆要求员工态度是最重要的，只有客人想不到，没有做不到。

康健一认为，虽然现在还有一些会所在模仿健壹公馆的建筑和经营模式，但健壹公馆的可模仿性并不强，"因为健壹公馆的核心文化是一种精神层面的东西，不是技术所能实现的。你可以刻意去模仿，但是你只能模仿一个局部，难以完全复制。况且，健壹公馆始终在不断超越自我，当你真正模仿好了的时候，我们又做得更好了。当许多人都跟在我们后面跑的时候，说明我把健壹公馆做成功了。"

当初选择这个风格的时候，康健一没有考虑过商业回报。他自嘲说，"当你什么都没有考虑的时候就投入的话，'傻瓜'都不足以形容了。"然而健壹公馆的成功，

雕塑：生旦净丑

让他更加相信："越不想谈商业，越不想谈回报，然后把事情做到极致，这样可能商业给你的回报反而会更多。"都说什么叫文化，其实把一件事情做到极致就叫"文化"。

志同道合

工作的性质决定了康健一需要花很多时间——差不多一年里有大半年——是在世界各地行走。但是他却说自己"不太会玩儿"。即使是看似在玩儿的过程当中，他多一半的时间也是在工作。他说他也特别渴望去真正放松、度假，什么都不想。"但是到目前为止，没有过。"

因为他喜欢建筑，喜欢艺术和文化，所以康健一特别喜欢并且善于观察。在度假的时候，他脑子也都想着这石材用得很巧妙，这地方设计很精致，人家服务员的服装很有特色，应该吸收一下。他还注意世界不同国家各个酒店独特的服务细节以及精神层面的一些理念。

"真正好的酒店你不会见到有形的服务，只是在你需要的时候及时有人会出现在你身边，这是人员的素质也是态度的问题，是值得我们学习的。"他说。

康健一更注重的是寻找志同道合者。他说，"物欲横流的世界里，能坚持做一件事的人，太少，坚持下来且做到优秀，更是难得。"

尽管难以寻觅，康健一还是找到了一位同"道"长者，他就是清宫御厨传人王希富老先生。当时康健一在大宅门酒店成功的基础上，希望更上层楼，便将目光投向了只有古都北京才能传承下来的宫廷菜和与之密切相关的庄馆菜。但是，宫廷已经消失了将近一个世纪，就连当年得了清宫御厨真传的老北京的八大楼厨师，也踪迹难寻了。康健一寻觅了很长时间，终于得以结识王希富老人。王老的外祖父陈光寿是前清光绪皇帝御厨的副庖长，父亲王殿臣师承其岳父，是20世纪二三十年代北京最享盛名的致美楼的掌灶厨师。王老本人家学渊源，

11岁就会做宫廷菜，几百道宫廷菜都在他脑子里，从起锅出锅及所有的原材料，都能讲得清清楚楚。

头一次见王老，康健一拿典型的老北京炸酱面作为主食来款待。面条端上来，王老一看就知道碗里的面条是哪一种，上不上档次。他说，面条有六种：头路条、二路条、葛条、荆条、韭菜扁、龙须面，你们给我上的这二路条，是给拉洋车和练摔跤的人吃的，不是宅门里给客人吃的。对炸酱面所用的酱，王老也不讲情面地批评：宫廷炸酱，酱得是稀黄酱，肉是带皮五花肉，是肉干炸而不是酱干炸，这酱太干了，能拌开吗？

虽然受到了批评，康健一却很高兴，因为他知道自己遇到了行家。于是他将王老请了过来，专门给健壹公馆的厨师做培训。王老对康健一挖掘并向世界展示中国文化的作为也非常欣赏，不要任何报酬也愿意为健壹公馆做顾问。如今年逾古稀的王老和小他几十岁的康健一，已经堪称忘年交。

有这样的志同道合者，康健一的企业跻身于世界精品酒店行业的高端，应该是理所当然的结果。

注重细节

在中国，很多想成大事者常常不拘小节。而康健一却偏偏是很讲究细节的人。他说，"挂的画要有点偏差我都会凭感觉知道。"有一次，他凭肉眼看出卫生间挂的一幅画有点偏，用尺子一量，误差两毫米，后来发现是绳子松了的问题。

这个自称"追求精细的人"感到最骄傲的，是他公馆的厨房。"厨房开餐的时候地板是干的，非常卫生干净。菜的出品和口感都非常精致。"他说。他未来的目标是做开放式的中餐厨房，"请最尊贵的客人一定要到厨房里摆一桌吃一餐，当着他的面做饭给他吃。"

注重细节，却绝不拘泥形式。康健一要求自己企业的宫廷菜必须中规中矩地遵照王老的嘱咐，食材必须要新鲜，时间一定要按照工序来做，绝不能偷工减料。但

健壹景园宴会厅外休息区

是对于宫廷菜的呈现方式，他却有自己的主张，把效仿皇家官窑烧制的餐具换成了具有现代感的白瓷盘。

"宫廷菜原汁原味，制作工艺不能改变，但呈现方式要改变，不能再用粉彩、青花。"康健一说，"中国的东西太讲究形式，形式大于用途，但不符合当今生活方式的改变。所以在我的酒店里，看到的都是东方的，享受的都是西方的。"

康健一本人追求把事情做到极致，注重细节，但是对他的管理团队和下属，却不会吹毛求疵。"我没有要求我的员工做到最好，我只要求他们比同行业人做得好一点点。"他说。这个"好一点点"，其实正体现了一种真正于细微处见精神的大智慧。因为，"服务、环境、菜品，都好一点点，客人不来，都没有道理了。"

健壹公馆的管理团队成员严谨、认真，都拥有海外的生活经历，拥有专业的酒店教育背景。康健一对于老板与职业经理人的职责区别，看得很清晰。他说，"老板就应该做老板的事，经营管理的事情由管理团队做决定，我基本不问不管。当我有不同想法的时候，我从不说'不'，我说：'请教你一下，这个问题是怎么考虑的？'当他们提出不同建议的时候，我不同意，但也不说'不'，我说：'让我考虑一下。'他们也不说'不'，他们永远说：'老板，你需要我帮助你做些什么？'只有这样的环境，才能让企业非常和谐。"

这"好一点点"的要求，让健壹集团旗下的所有项目都良性运转。就拿他的厨师团队来说，"他们天天看书学习厨艺，研究厨艺，他们很有责任心，自觉地研发新菜。"康健一欣喜地说，"当一个厨师都在看书，我觉得我真的赢了。"

魅力与分配

康健一认为，一个企业成功的关键在于两点，一个是老板的个人魅力，另一个就是分配。

信奉佛教的康健一和夫人拜藏传佛教萨迦派高僧为师，每日修行，"得失心不那么重"，他说。心态平和，又有深厚的文化底蕴，使很多人觉得这个自称"不招人烦"的企业家谈吐儒雅，却自有一种强大的气场。这种气质，自然会赢得下属的敬服。

而分配方面，康健一认为十分重要的是把利益最大化地分配给所有的人。健壹集团的员工除工资、奖金，每三个月还有一次分红，让最基层的员工都有相当不错的保障。优良的工作环境也会给员工带来很大的自信，不仅能够得到别人的认可，还能得到相对高的回报，谁还会不用心工作呢？

"员工都在说你好，你不好都不行了。"康健一说。持续良性的分配，重金吸纳人才，尽管回收慢些，但根基是扎实稳固的，这比盖一栋楼、挣大笔钱更让他有成就感。"别人家是找不到人，我们家是从来不缺人。"他说。在人才流动率超高的酒店和餐饮业，年平均流失率不超过30%的就算好企业，健壹的流动率保持在3%。"这是健壹最大的成功。"

所以，康健一会说，"生意不好，永远不是市场的问题，你把所有的都准备好了，把最好的东西呈现了，把产品、消费习惯都研究透了，生意怎会不好？"

现在，他的时间，只放在一件事情上，那就是向全世界"推销"健壹公馆以及中国传统建筑文化与饮食文化的博大精深。"每个人，不是有钱有权或者美貌倾城才是成功。"他说，"在历史的长河中，这些都只是幻影。伟大，一定是永恒的东西。"

Pursuing Perfection
About Kang Jianyi

Every year on the first day of the Spring Festival, he will pay a visit to the Forbidden City. While strolling through the walls of the palace, he envisions creating a replica of Taihe Hall on the lawn of JE Mansion. This is the personal New Year habit of JE Mansion's chairman, Kang Jianyi.

As a born-and-raised Beijinger, Kang has traditional culture, art and history flowing through his veins. Even his name is infused with traditional Chinese culture. The name chosen by his parents was Kang Jian, the two characters forming the word "health" in Chinese, wishing him physical and spiritual health. In 2003, he added an additional character, that meaning "one", which both represents the Daoist concept that all matter in the world begins from "one", and also indicates the philosophy that all those of strong morals are of the same heart.

After studying Chinese painting, and later working in auctioning and collecting, in 2001 Kang began dedicating himself to the hotel dining industry. He made this decision simply because at the time he "liked it", but admitted that it was also "meant to be". He explains, "My wife's family is in the catering industry, and it was only a matter of time before I got into it." Although it may have seemed like a rather arbitrary decision, Kang's obsessive adoration for traditional culture ensured that his hotel would have a cultural quality like no other. After opening the Dazhaimen Yingxiang International Business Hotel, he decided to establish another hotel, this time with a much stronger traditional Chinese ambiance, JE Mansion.

During the past century of the hotel industry in China, especially after the late 1970s when China began cooperating in joint ventures, high-end luxury hotels have become increasingly commonplace. However, Kang discovered the mutual shortcoming of these hotels: "When people go to France they want to see castles and The Louvre; likewise, when people come to China they want to see places that are distinctly Chinese." But among all the luxurious hotels in China, even garden-style ones, there is very little of traditional Chinese culture to be seen, as the goal of the vast majority of hotels is merely to gain profit.

Kang, on the other hand, with his infatuation for history and culture, treats the hotel industry as if it were the culture

President of JE Group: Kang Jianyi

industry. Although he has had success in auctioning, and has successfully established several upscale estates named after himself, the chairman of JE Group does not consider himself a businessman. He says that he is not doing business, he's making art.

Just like many artists, Kang establishes hotels and estates purely to do what he "wants to do", and seldom considers the ratio of investment and return. Based on his many years of experience in hotel management, from the early planning stages until present, he says that he has not "run into much trouble or difficulty" and that "everything has gone quite smoothly". There is only one point he refuses to change: the flawless combination of old and new, Chinese and western culture and art with modern practicality, to create a hotel that will make the Chinese proud, and earn the respect of visitors from abroad.

This is Kang's dream. But this is not only a dream to indulge in, he has put this dream into motion. Kang says, "JE Mansion expresses my insight toward art and my comprehension of traditional Chinese culture, and I hope it will act as a medium to promote high-quality cultural exchange between China and the west."

When stepping into JE Mansion, one feels as though they have stepped into a museum. In the main hall a selection of Qing Dynasty emperor's robes is displayed, rustic brick and stone carvings decorate the subtle yet elegant stone walls and wooden floors, and in the same room furniture from late feudal China, many from Kang's personal antique collection, can be found alongside contemporary European sofas. What is especially unique is the estate's collection of antique doors, hundreds of doors procured from throughout China, some only worth about two thousand yuan, some valued at tens of thousands. "Each door has its own story, each has witnessed its own period of history and watched over its own family," Kang says. "Other works of art are to be looked at, this kind can be used. Nowadays there are lots of vintage restaurants in Beijing, and some have old decorative doors, but here each room is equipped with an antique door, and all of them are actually used on a daily basis. I take what other people have no use and make it into something valuable."

The museum-like atmosphere and collection of doors are among the many ways in which Kang has striven to combine culture, art and modern practical living needs. "There's no sense in putting artifacts in a storehouse where no one will see them, only by sharing them with everyone can you fulfill their true value," Kang says. "Things in museums can be viewed but not touched, whereas everything we have here can be both looked at and used."

JE Mansion is a museum which one can "experience".

One of a Kind

A successful operation and management scheme is certain to be imitated and reproduced by other. Kang is very confident about this matter, welcoming others to try their best. "We're not worried about being imitated. A successful business has its own soul, the things which make up its essence can't be copied; even if someone makes something really similar, there will still be lots of details that just aren't the same."

His confidence stems from a particular concept: as a service enterprise, the essence of JE Mansion is not just in its architecture, more important is its service philosophy and attitude, which will give customers a truly one-of-a-kind experience.

Kang believes in striving for exceptionality and absolute uniqueness, and JE Group's hotels reflect this pursuit. For example, a completely traditional Chinese environment

JE Mansion guest room

would not be comfortable enough, but an entirely western decor lacks Chinese culture; so JE Mansion combines the two styles of living, from architecture to display and service, and Chinese culture can be seen everywhere, but combined with convenient modern living facilities.

At the beginning of Kang's endeavors he believed that the most appealing aspect of a hotel was its architecture; but with the development of JE Mansion and the broadening of his own horizons, Kang gradually realized that JE Mansion couldn't be compared to any building from elsewhere in the world. The culture of each people results in their own local architectural characteristics, and it would be wrong to say which is better than the others.

"Creating a fine Chinese-style hotel is not a matter of simply piling a bunch of Chinese things together, what's more important is bringing out the rich culture which can't necessarily be seen on the surface", says Kang. Works of art are scattered throughout JE Mansion, in an intentional slightly disordered manner, making it feel more like a home. "The word mansion comes from the Latin 'mansio', which means 'dwelling', and here everything I like is placed all over to share with others", says Kang. "I hope that when people visit here they can make themselves at home, and feel the warmth of home here. Hotels are for 'dwelling' in, not for looking at."

Based on this principle, although Kang has continued to

promote Chinese culture, in recent times he has done so in a more and more simple manner. "I've learned that simplicity is most efficient, just doing things as simple as possible; I used to add so much, but lose the focus," he says.

After several years of fine-tuning, JE Mansion has become its own unique self. First, in terms of dining, JE Mansion features dishes of impeccable quality based on imperial Chinese cuisine. The hotel has joined RELAIS & CHATEAUX, the world's highest-level hotel industry alliance, as one of only three hotels throughout China able to meet the strict dining standards of the organization. Second, JE Mansion's architecture and atmosphere are irreplaceable, a tranquil setting which is truly difficult to come by among the hustle and bustle of the city. Most important of all is the staff's service; anything the guest can possibly need, they will find a way to provide.

Kang believes that although there is a handful of hotels and clubs mimicking JE Mansion's architectural and managerial style, none have done so very accurately; this is because, "the central culture of JE Mansion lies in its spiritual essence, which cannot be achieved with any kind of technology. You can imitate it, but you can only mimic a part of it, and copying the entire thing would be really difficult. Also, JE Mansion is always undergoing self-improvement, so even if you get your hotel to be exactly the same, by that time we'll already have made ours better. When I see that there are

JE Mansion's 50,000 sq. m lawn

a lot of people trying to catch up with us, which means that JE Mansion is successful."

When he first chose the hotel's style, Kang did not pay much attention to commercial gain. He says, self-depreciatively, "When you do something big without thinking about it at all, 'stupid' fails to fully describe the magnitude of the situation." But JE Mansion's success makes him believe more clearly that, "Actually, the less you think about business and profit, and just want to make things great, the more success you'll have." Everyone discusses the definition of culture, but in fact culture is merely dedication to perfection.

A Common Vision

The line of work Kang has chosen as resulted in the fact that he spends much of his time, about half of each year, traveling around the world. But he claims that he "isn't much of a fun-loving guy"; although it appears that he is traveling for pleasure, the majority his time abroad is spent working. He says that he really wants to take a relaxing vacation, but, as of yet, he hasn't "had the chance".

Because of his love for architecture, art and culture, Kang enjoys and is very skilled at observation. Even when supposedly on vacation, he will always be thinking about things like the clever stone arrangement of a certain building, or the fine design or another, or the interesting staff uniforms. He also pays attention to the unique details and principles of hotel service in different countries. "At a truly great hotel service isn't something you can perceive, it's something that will be there the moment you need it, it's a matter of the staff's qualities and attitude, and is something we should strive for at our own hotel," he says.

Kang has always been searching for other people with similar ambitions and aspirations. He says, "In a world where things are always changing so quickly, it's hard to find someone willing to stick to doing one thing, and even harder to find someone who does it well."

Although it's difficult to find, Kang has in fact come across someone on the same path as himself, Mr. Wang Xifu, a successor of royal Qing cuisine. Once the Dazhaimen Hotel had attained a certain level of success, Kang hoped to take the hotel to the next level, and focused his attention on royal cuisine, and the closely related zhuangguan cuisine, both of which can only be found in Beijing, the capital of the Qing Dynasty. However, feudal China has been abolished for almost a century, and even the knowledge passed on by master chefs, such as those from the "eight great restaurants" of old Beijing, is all but lost. Kang searched all over, and finally crossed paths with Mr. Wang. Mr. Wang's grandfather Chen Guangshou was the second head chef of Emperor Guangxu, and his father Wang Dianchen was the head chef of the most famous restaurant in Beijing during the 1920s and 1930s. Wang followed in his grandfather and father's footsteps, and was able to cook royal dishes by the age of 11. He has several hundred of these recipes stored in his memory, which he can recall with encyclopedic accuracy.

The first time he met Mr. Wang, Kang had prepared a typical Beijing dish, zhajiangmian (noodles with soybean paste sauce), for his guests. After the noodles were placed on the table, Mr. Wang began to critique the quality of the noodles, saying that of the six types of noodles these were of the erlutiao variety, which is served to rickshaw runners and wrestlers, not to guests at one's home. He also criticized the sauce Kang had used, saying that for royal soybean paste sauce it has to contain pork, with some skin and fat, and it's the meat that's dry deep-fried not the sauce, pointing out that Kang's sauce was impossible to mix together with the noodles.

Although he had been criticized by Mr. Wang, Kang was extremely pleased, because he knew he had met the person he was looking for. So he invited Wang to train his team of chefs at JE Mansion. Mr. Wang also highly admired Kang's desire to research and promote Chinese culture, and agreed to become JE Mansion's cuisine adviser, free of charge. Thus despite of their age difference, Kang being several decades younger than the elderly Mr. Wang, the two began a deep and lasting friendship.

Only by sharing a similar vision with a friend possessing such rare skills and knowledge was Kang able to establish the status of JE Mansion among the finest hotels in the world.

Attention to Detail

In China, many people want to do big things, but fail to pay attention to small details. Kang, on the other hand, is particularly attentive of details. He says, "If a painting is slightly slanted I can just feel that something is not right." One time, he saw with his bare eyes that a painting in the washroom was crooked, so the staff measured the painting with a ruler, and it was off by two millimeters, which was discovered to be the result of the string having slackened.

A self-proclaimed "meticulosity fanatic", what Kang is most proud of is JE Mansion's kitchen. "When meals are served the kitchen floor is dry, very clean. The dishes both look and taste extremely exquisite," he says. His future goal is to create an open-style Chinese kitchen, to invite distinguished guests to join him in the kitchen, and prepare a meal for them.

Although paying attention to detail is important, the form itself cannot be ignored either. Kang requires that all royal dishes are made strictly according to the instructions of Mr. Wang, the ingredients must be fresh, and all the proper steps must be taken, with absolutely no shortcuts. But Kang also has his own way of serving royal dishes, replacing the imitation royal kiln ware with simple modern white porcelain plates.

"For royal cuisine to be authentic the production process must stay the same, but the way of serving can change; we have to stop using pastels and blue and white porcelain," Kang says. "Chinese things tend to put too much focus on the form, the form becomes more important than the use, but this isn't practical for modern life. So at my hotel, everything you see is Chinese, but everything you use is western."

Kang strives for exquisiteness and pays much attention to detail, but with his management team and staff he is never unfairly critical or nitpicky. "I don't ask them to do a perfect job, just slightly better than others in the same industry", he says. This "slightly better" is an example of Kang's subtle yet management philosophy, because "if the service, environment and food are all slightly better, then the customers will come to our restaurant".

JE Mansion's management team are strict and diligent, all have professional restaurant education backgrounds, and experience living and working abroad. Kang makes a clear distinction between the responsibilities of an owner and a manager. He says, "An owner should do an owner's job, any operational and management decisions are made by the management team, so I seldom ask what they're up to or tell them how to do things. When I see things differently than them I never say no, I say, 'What do you think of this situation?' When I disagree on something, I just say, 'Let me think about it.' They never say no either, they always say, 'Boss, what can I do for you?' Only in a harmonious environment like this can a business run smoothly."

This "slightly better" request has ensured that all of JE Group's projects operate smoothly. Take his kitchen team for example, "Every day the chefs studying and researching the culinary arts, they have a very strong sense of responsibility, and actively create new dishes on their own," Kang is pleased to say. "When my chefs are all busy reading, I really feel like I've won."

Charm and Allocation

Kang believes that the success of an enterprise is based on two aspects, one is the personal charm of the owner, and the other is allocation of funds.

Kang and his wife, both followers of Buddhism, have been accepted as disciples of a high Sakya lama monk, and practice every day. "Gain and loss are not all that important," Kang says. He has a calm state of mind and a deep cultural background; this causes many to find his rather refined way of talking about business matters, in his own words, "not annoying". However, Kang still has a kind of strong charismatic aura about him, which has naturally earned him the respect of many of his staff.

As for allocation of funds, Kang believes that it is very important to divide the larger portion of profits among everyone involved. In addition to salary and regular bonuses, every three months JE Group employees receive a share out bonus, allowing all staff to have a strong sense of security. A great work environment also gives the staff confidence, as by working hard not only are their efforts recognized by others, but they will also receive an equally high monetary reward.

"If all the staff think you're great, then you must be," Kang says. High pay and continuous bonuses attract talented staff. Although it may take longer, Kang finds his business more satisfying than constructing buildings and making money the fast way. "Other places have trouble finding staff, but we're never short on people," he says. In the restaurant and dining industries, if the staff turnover rate can be controlled within 30% it's considered quite low, but at JE the rate is less than 3%. "This is JE's greatest success."

Kang explains, "So if business is bad, it's never a problem with the market; if you prepare everything right, offer your customer the best products, and learn everything there is to learn about consumer habits, how would business still be poor?"

Currently, Kang devotes his time to one matter alone, namely "promoting" JE Mansion and traditional Chinese architecture and cuisine culture to the world. "Money, power and good looks do not denote success," he says. "In the long run, these are just illusions. True greatness is eternal."

The Mansion Gate

健壹公館菜・小食

风味小食分享
Desserts

玫瑰饼

玫瑰花中含有丰富的多酚类和黄酮物质,有抗氧化作用,可清除色素沉着,有美容、养颜、排毒、调节机理之功效。口味特点:玫瑰香味浓郁,甜香适口,口感松软。

Rose Cake

A crispy pastry, rose cake, or Meigui Bing, is stuffed with roses picked fresh at the Miaofeng Mountain in the western outskirt of Beijing before preserved and made into jam. It is a delicacy for spring time, and has inherited a Chinese tradition of eating fresh flowers since ancient times.

Ingredients include cooked cake flour, lard, rose jam, shelled pine nut and sesame.

原　料

油酥:低筋粉500克,猪油250克;
水皮:低筋粉500克,猪油125克,糖20克,水350克;
玫瑰馅800克,松子20克,芝麻20克,熟面粉适量。

制作方法

1. 将500克面粉加猪油250克搓成油芯。
2. 将500克面粉加猪油125克、糖20克、水350克和成水油皮。
3. 将油芯包入水油皮中,按扁擀开叠成三折。再擀开卷起,揪成每个剂15克待用。
4. 在玫瑰酱馅中加入白芝麻、松仁、熟面粉一起搅拌成馅。
5. 将剂子擀开包入15克玫瑰馅,按扁成饼状。用上火180℃下火205℃烤16分钟即可。

备注:开酥手劲一定要均匀,否则层次不够分明。

健壹公馆菜·小食

奶酪果子冰

奶酪果子冰是一道传统的中式宫廷甜品。中国人对于冰沙的食用历史很早。奶酪香甜适口，既有淡淡的酒酿的清新气息，又有杏仁和葡萄干的香气，整个甜品层次分明，相互协调，草莓冰沙的酸甜使得奶酪更加香甜。

Cheese Icy Fruits

A traditional dessert on the royal menus of the Qing Dynasty, cheese icy fruits, or Nailao Guozibing, is a refined upgrade of fermented glutinous rice wine. Fusing the flavor of sweet and sour, it should taste cool.

Ingredients include shelled fresh walnut, sliced apricot, raisin, fermented glutinous rice wine, fresh milk, ice strawberry ball, sugar and mint leaf.

原　　料

鲜核桃仁1粒，杏片2片，葡萄干1颗，醪糟10克，鲜牛奶20克，草莓冰球15克，白糖7克，薄荷叶1片。

制作方法

1. 准备好小碗，在小碗里放入鲜核桃仁、葡萄干、杏仁。
2. 把白糖放入牛奶中，冲入醪糟中混合，再倒入小碗上蒸炉，蒸15分钟。出锅放凉，自然凝结成冻。
3. 将草莓冰球放置在蒸好的奶酪上，点缀薄荷叶即可。

备注：蒸奶酪时蒸锅中途不可以打开，否则奶酪会凝结不好。

杏仁豆腐

此小吃色白光洁，比嫩豆腐还要细嫩，浮于清澈的糖水上面，甘冽醇厚，沁人心脾，为盛夏消暑的佳品。

Almond Jelly

Known as *Xingren Tofu*, or almond bean curd, in Chinese, almond jelly has nothing to do with soybean. It got the name just because the apricot milk is white and tender, resembling the most refined bean curd. It has been a refreshing dessert for summer.

Ingredients include extract from southern or northern apricot kernel slices, milk, agar, xylitol and sugar.

原　　料

杏仁片榨汁1000克、牛奶250克、白糖100克、琼脂80克、纯净水适量

制作方法

1. 杏仁片榨成汁，1000克杏仁片加水3000克，过箩去掉渣。
2. 杏汁加入牛奶、白糖、琼脂熬开锅3分钟装入小碗里，凉后放入冰箱冷藏，凝结成半固体的杏仁豆腐。
3. 纯净水化开白糖倒入碗中，吃时可撒鲜果丁。

备注：控制好软硬度才能保证口感细腻。

奶卷子

奶卷子是蒙古族地区百姓的主食，皮色乳黄，质地软嫩细润，馅为红白两色，酸甜醇香，入口即化，奶香浓郁，营养丰富。

Yogurt Roll

Originally from Inner Mongolia, yogurt roll, or Nai Juanzi, used to be indispensable for Mongols as rice for people south of the Yangtze River. It is considered creative to use yogurt as wrapping and the typical roll would melt at the very bite.

Ingredients include milk, sugar and haw jelly. The milk has to be boiled over low heat till it thickens and the cooled milk skin is made into the wrapping to roll haw jelly inside.

原料

牛奶2500克，白糖200克，山楂糕200克，芝麻200克。

制作方法

1. 将牛奶熬制浓稠后晾凉。
2. 将晾好的牛奶，用油纸做成奶皮，把芝麻白糖馅和山楂糕馅分别横摊在奶皮上，各占一半的面积，再垫着油纸将前后两端分别向中间卷成如意形，切小块即可。

备注：煮牛奶时掌握好时间，没煮好会影响奶皮的光滑和软嫩。

JE Mansion Cuisine *Desserts*

豌豆黄

豌豆黄是北京的传统小吃，曾同芸豆卷一起传入清宫成为御膳的名点。其颜色浅黄，细腻凉甜，入口即化，是夏季消暑佳品。曾有文人雅士赞誉："从来食物属燕京，豌豆黄儿久著名，红枣都嵌金屑里，十文一块买黄琼。"

Pea Cake

A traditional snack of Beijing, pea cake, or Wandou Huang, together with kidney bean roll became popular dim sums of the royal cuisine during the Qing Dynasty. In yellowish color, pea cake is refined, cool and sweet, melted at the very bite. It is among the delicacies for summer, and has been praised as yellow jade.

Ingredients include shelled peas, jujube, sugar, and agar. The shelled peas have to be soaked overnight before being steamed and sieved into pea puree.

原　　料

去皮豌豆1000克，白糖180克，水适量。

制作方法

1. 将干豌豆挑去杂质在清水里冲洗干净，加凉水将豌豆浸泡一夜，隔天早上将豌豆控干水分，放托盘铺平，再加没过豌豆的水，放入蒸箱蒸40分钟，用细箩过箩成泥。
2. 将成泥的豌豆黄、白糖倒入锅内，小火熬至浓稠状态，出锅入盘，盖上油纸放凉。吃时根据喜好切块或用模具刻出造型。

备注：熬制豌豆黄时火不要太大，否则豌豆泥容易飞溅烫伤人，也容易糊锅；熬豆不宜用铁锅，因为豌豆遇铁器容易变成黑色。

健壹公館菜・小食

芸豆卷

传说清代慈禧有一次偶然听见宫外有人打铜锣叫卖芸豆卷，一时高兴，把那人叫进宫来，吃了他做的芸豆卷觉得很好，便把那人留在宫中专为她做吃食。芸豆卷口感细腻，芸豆香味突出。

Kidney Bean Roll

Originally a street dessert consumed by common people, kidney bean roll, or Yundou Juan, found its way onto the imperial dinner tables during the Qing Dynasty and became a favorite of Empress Dowager Cixi.

Ingredients include white kidney beans, xylitol and red bean paste. Preparation involves soaking the shelled kidney beans, crushing them, boiling and steaming. The kidney bean paste should be neither too hard nor too soft.

原 料

白芸豆1000克，白糖300克，红豆沙适量。

制作方法

1. 将芸豆用水泡涨去皮，用水清洗干净，再用开水锅余烫5分钟，控干水分放在铺有布的蒸笼上，干蒸40分钟拿出晾凉。
2. 用压面机压成泥状，加上适量的白糖拌匀，用布垫着擀成2毫米的长方形，抹上1毫米的红豆沙，再卷成如意卷改成2厘米的段即可。

备注：芸豆面软硬度要合适，太干时可加纯净水少许。

驴打滚

驴打滚是北京的代表食品之一,也是各种庙会经营的应时小吃。北京专门写京俗的《竹枝词》《燕都杂咏》中都写有"驴打滚"的内容:"红糖水汁巧安排,黄面成团豆面埋,何事群呼驴打滚,称名未免近诙谐。"

Soybean Cake

A typical Beijing snack, soybean cake is known as Lv Da Gun, or Donkey Roll About, in Chinese. It was so named because the glutinous rice cake should be rolled in the soybean flour before it is cooked, as if a donkey rolls about in the wild.

Ingredients include glutinous rice flour, soybean flour, xylitol, and brown sugar water.

原　　料

糯米粉500克,黄豆面550克,白糖200克,水700克,红糖水适量。

制作方法

1. 将黄豆面用200℃炉温烤至深黄色,直到烤出黄豆的香味放凉过箩。拿出同等份量的黄豆面粉跟同等份量的白糖一起混合当馅。
2. 准备一个托盘,刷上薄油,将糯米粉、糖、水一起混拌均匀过箩到准备好的托盘里,上笼蒸20分钟至熟为止,拿出来刷上薄油晾凉。
3. 用黄豆面当扑面把蒸好的糯米面擀成2.5毫米的厚片,撒上带糖的黄豆面,顺一个方向卷起来成大拇指粗,再改成2厘米的小段。吃时蘸红糖水即可。

备注:蒸熟后等表面水份散发掉再刷油,可保证其口感韧性十足。

JE Mansion Cuisine *Desserts*

苏子茶食

苏子茶食是喝茶时配的一种中式点心。苏子味干香浓郁,香气独特。苏子有润肠、止咳、平喘的功效。

Perilla Seed Cake for Tea

Century-old perilla seed cake, or Suzi Chashi, is a typical Chinese dim sum served with tea.

Ingredients include cooked cake flour, lard, black perilla seeds, white perilla seeds, sesame, salt, sugar and sesame oil.

原　　料(70个,每个30克)

水皮:低筋粉500克,猪油125克,糖20克,水350克。

油酥:低筋粉500克,猪油250克。

苏子馅:黑苏子100克,白苏子100克,芝麻20克,盐8克,香油30克,糖20,猪油40克。

制作方法

1. 将苏子烤熟取100克压碎,100克压饼状,加入盐、糖、香油、猪油、芝麻一起拌匀,口味是咸甜味。
2. 将500克面粉加猪油250克搓成油芯。
3. 将500克面粉加猪油125克、糖20克、水350克和成水油皮。
4. 将油芯包入水油皮中,按扁擀开成三折。再擀开卷起,揪成每个剂15克待用。
5. 将剂子擀开包入15克苏子馅,按扁成饼状,上火180℃下火205℃烤制16分钟即可。

备注:苏子烤时要控制好火候,出香味即可。

JE Mansion Cuisine *Desserts*

果子干

果子干是一款地道的北京甜点,立春前食用最佳。据传1866年即同治皇帝9岁那年,他私自走出故宫直至西安门街头吃果子干,一时传为佳话。此品甜酸适度,软绵利口。

Dried Fruit Mix

A typical dessert of Beijing, dried fruit mix, or Guozi Gan, is best served before the Spring Begins, the first of the 24 solar terms on the Chinese lunar calendar, which usually falls in early February.

Ingredients include dried persimmon, Beijing red pear, dried apricot, lotus root and sugar. Dried persimmon and apricot have to be soaked and chopped in the preparation of this snack.

原　　料

柿饼250克,红肖梨45克,纯净水250克,杏干60克,莲藕35克,柠檬酸5克,白糖125克。

制作方法

1. 将柿饼去蒂撕碎,杏干用水洗净、泡软,藕刮去皮、洗净切片,用开水煮熟备用。红肖梨去皮切片。
2. 将柿饼碎加入泡过杏干的水并用筷子搅动,直到液体具有粘稠性即可。加入白糖和柠檬酸。
3. 在碗中倒入柿子水,再放入两颗柿饼碎、两片杏干、两片红肖梨,面上放入藕片,入冰箱即可。随吃随取。

备注:柿饼选用肉厚的,否则搅拌完后就只剩皮了。

萨其马

萨其马因其满族糕点制作手法保存比较完整，所以被称为满族糕点的活化石，是北京著名京式糕点之一。萨其马色泽米黄，口感酥松绵软，入口即化，香甜可口，桂花蜂蜜香味浓郁，深受食客的喜爱。

Caramel Treats

A typical snack of Manchu, caramel treats, or Sachima (also spelled Saqima), used to be one of the ritual foods for the founding fathers of the Qing Dynasty. A sweet snack, it mainly consists of flour, butter, and cerealose or caramel, and is popular in Beijing all year round.

Ingredients include bread flour, baking powder, cerealose, vanilla powder, eggs, haw jelly, green plum and cooked sesame.

原 料

富强粉500克，臭粉6克，饴糖750克，白糖150克，鸡蛋400克，山楂糕50克、青梅50克、熟芝麻20克。

制作方法

1. 将富强粉、鸡蛋、臭粉和在一起成面团，醒大约10分钟，然后擀制成薄厚一致约0.5厘米的面片，切成大约10厘米长、韭菜叶宽的条。
2. 放入四成油温的油锅中，炸成乳黄色。
3. 把饴糖和白糖加水熬制成糖浆，倒入炸好的面条上，进行快速搅拌，拌匀为止，然后放入铺好熟芝麻的不锈钢盘中，压实，上面撒上山楂糕、青梅点缀，晾凉切块即可。

备注：在熬制糖浆时，火候是影响糖浆软硬粘稠度的重要因素，搅拌时一定要快，避免糖浆烫手。

芙蓉糕

芙蓉糕是在萨其马的基础上，加入多一点的果料和白糖、苋菜汁的装饰和调味，色彩更加艳丽，口味层次更加饱满。

Lotus Cake

A variation of caramel treats, lotus cake, or Furong Gao, got the name because it's in the color of lotus flower.

Ingredients include cooked cake flour, bread flour, baking powder, honey, eggs, syrup, green plum, haw jelly, raisin, cooked sesame and sugar.

原　料

低筋粉250克，高筋粉250克，泡打粉10克，蜂蜜300克，鸡蛋5~6个，糖浆300克，另备装饰用青梅30克、山楂糕30克、葡萄干20克、熟芝麻20克、白糖80克、青红丝30克、枸杞子20克。

制作方法

1. 将所有配料和在一起稍醒20分钟，然后进行擀制，可用淀粉做扑面，擀至薄厚约0.5厘米的片，切成长10厘米的面条。
2. 放入四成油温的油锅中，炸成乳黄色。
3. 把蜂蜜加糖浆熬至用手指粘有丝状，倒入炸好的面条上进行快速搅拌，拌匀为止。然后放入铺有青梅、葡萄干、山楂糕、枸杞子、芝麻的盘中，压实。上面撒上装饰料点缀、晾凉、切块，撒上白糖，淋上苋菜汁，再撒上青梅即可。

备注：在熬制糖浆时火候要掌握好，它是决定软硬的关键，搅拌时要快。

元宵

元宵也是节令性食品,农历正月十五食用。唐开元年间就有"家制米圆相饷,即呼之为元宵"的记载。元宵本称"圆子",也叫"汤圆"。古代农历正月十五日是"上元节",这天晚上称"元宵"或"元夜"。按古代风俗,上元节晚上总要吃圆子,以取家庭团圆幸福之意,后来人们逐称"圆子"为"元宵",相沿至今。

原　　料

A:　糯米粉500克,澄面100克,糖100克,开水适量。

B:　猪油50克,黑芝麻100克,糖粉400克,水适量。

制作方法

1. 将黑芝麻烤熟,晾凉后压碎,与糖粉、糯米粉一块拌匀,加少许水,装进1厘米厚的托盘,压平,隔日将芝麻馅切成1厘米见方块状。
2. 笸箩里放上干的糯米粉,把切好的芝麻馅放在清水中蘸一下,再放回到糯米粉中摇滚,等均裹上糯米粉。裹不上糯米粉的再拿出来蘸水放回糯米粉中摇,反复几次直至裹到直径3.5厘米左右即可。
3. 食用时,放在开水锅中煮至漂浮上来即可,带汤可加少许糖桂花,食用味道香甜软糯并有淡淡的桂花香气。也可用针在元宵表面扎眼炸着吃,外酥里软香甜可口。

备注:煮元宵的火不要太大,不然会造成熟透的元宵外形破损。

Glutinous Rice Ball

A typical food for the Lantern Festival on the 15th day of the first month on the Chinese lunar calendar, glutinous rice ball, or *Yuanxiao* or *Tangyuan*, already prevailed during the Tang Dynasty more than 1,200 years ago. The festival marks the end of the celebration of the Chinese lunar New Year and the rice balls are considered best suitable for the first full moon of the new year, which symbolizes family reunion according to Chinese customs.

Ingredients include glutinous rice flour, wheat starch, lard, black sesame and crystal sugar.

宫廷炸春卷

春卷是中国农历立春时的时令性小吃，用烙熟的圆形薄面皮卷裹馅料，成长条形。然后下油炸至金黄色浮起，即可捞出。而馅料可荤、可素、可咸、可甜。如韭黄肉丝春卷、荠菜春卷、豆沙春卷等。宫廷炸春卷皮酥香，馅鲜嫩。

Royal Fried Spring Roll

Another traditional food for spring, royal fried spring roll, or Gongting Zha Chunjuan, can have either meat or vegetarian fillings. It can be salty or sweet. The rolls are fried when the oil temperature reaches 180 degree Celsius.

Ingredients for royal fried spring roll include wheat flour, shredded pork, green bean sprout, hotbed chives, sesame oil, soy sauce, cooking wine, pepper powder, and peanut oil.

The dough for spring roll is kneaded with water and salt.

原　　料（20个）

面粉200克，猪肉丝100克，绿豆芽50克，韭黄50克，盐5克，香油5克，酱油6克，料酒4克，胡椒粉0.5克，花生油1000克。

制作方法

1. 将面粉加适量盐、清水和成软面团，反复揉透揉匀，抓上劲达到滋润不粘手为止，用一手提面团，一手转刷过油的饼铛，将面团往饼铛上一按，然后提起面团即可摊成直径约15厘米大小的春卷皮。
2. 将绿豆芽、韭黄洗净，韭黄切寸段。炒锅上火，放底油烧热，放葱花随下猪肉丝，煸炒断生入料酒、酱油入味，再下绿豆芽，出锅后拌之香油、盐、胡椒粉及韭黄成馅。
3. 用摊好的皮包适量的馅，卷成大拇指粗的卷，用面糊粘住接口即可，下油炸至金黄色，浮起即熟。
4. 春卷皮和面比例必须合适，做到面软而不流，做出的皮薄如蝉翼。

备注：油温升到180℃时方可下锅炸制，油温高易炸不熟，油温低炸制易窝油，成品干硬。

老北京炒合菜配春饼

春饼是汉族立春时的节令性食品。吃春饼也称之为"咬春",人们将春韭、银芽等炒在一起卷饼吃,饼筋斗,味香浓,春韭清香,银芽脆嫩。

Spring Pancake of Old Beijing

A specialty for the Spring Begins, the first of the 24 solar terms on the Chinese lunar calendar, which usually falls in early February, spring pancake or Chunbing of old Beijing, has been popular in the ancient capital. It is a thin pancake made of bread flour as wrapping, which rolls various sautéed shredded tender vegetables plus sauced pork and duck inside up. It differs with spring roll in that the wrapping is larger and the roll is not fried.

Ingredients for the spring pancake include bread flour, sauced pork, sauced duck meat, smoked pork, smoked chicken, donkey meat, sautéed shredded pork with hotbed chives, sautéed shredded pork with silk noodles, scrambled egg, sautéed shredded pork with bean sprout and sweet fermented flour sauce

原　　料
高筋面粉500克,开水225克,盐5克,色拉油适量。

配　　料
酱肉100克,酱鸭100克,熏肉100克,熏鸭100克,酱鸡100克,熏鸡100克,白肉100克,韭菜100克,粉丝50克,银芽50克,鸡蛋5个。

制作方法
1. 把开水分次加入面粉中,搅拌至没有干粉,揉成团,盖上湿布醒15分钟。
2. 醒好的面团揪成每个20克的剂子,按平,把色拉油刷在面皮上,撒少许干粉,扫掉再刷一层油,再撒少许干粉(反复2次即可)刷少许油,把两张面皮合在一起,稍按平备用。
3. 平底锅烧热,把擀好的面皮放入锅中,烙半分钟左右,待表皮起芝麻状时翻过来,再烙制半分钟,鼓起即可,分层揭开即成两张春饼皮。
4. 将白肉切成肉丝,韭菜切寸段,粉丝用刀斩一下,银芽去两头。肉丝放入锅中煸炒出香气时下银芽、粉丝、韭菜,调味翻炒均匀出锅装盘。
5. 鸡蛋磕开放入碗中打散,加入少许盐上锅摊成蛋饼,盖在炒好的合菜上即可。
6. 将春饼皮包上炒熟的馅料,成卷状即可食用。

备注:擀饼时要两面颠倒均匀地擀制,烙出饼花才会出来饼的香味。

JE Mansion Cuisine *Desserts*

北京粽子

农历五月初五是端午节,中国人以吃粽子来纪念战国时期的楚国大臣屈原。粽子多为三角形或斧头形,或枕头形。北京粽子个头较大,为斜四角形或三角形。目前市场上多为糯米粽。在农村,仍然习惯吃黄米粽,黏韧而清香,别具风味,多以红枣、豆沙做馅,也有果脯为馅。北京粽子以香甜为主,口感软糯。

Glutinous Rice Dumpling of Beijing

A special food for the Duanwu or Dragon Boat Festival, on the fifth day of the fifth month on the Chinese lunar calendar, glutinous rice dumpling, or Zongzi, is believed to have been created to commemorate the death of poet Qu Yuan (c. 340–278 BC) of the ancient state of Chu during the Warring States Period. A descendant of a royal family of Chu, Qu Yuan served in high offices but was demoted and exiled as he opposed to the king's decision to ally with the increasingly powerful state of Qin. He finally committed suicide by drowning himself in the *Miluo* River in present Hunan Province on the fifth day of the fifth lunar month. The local people who admired him dropped glutinous rice dumplings wrapped in bamboo leaves into the river to feed the fish so that they would not eat Qu Yuan's body. Making and eating pyramid-shaped glutinous rice dumplings have since become Chinese folk customs. Glutinous rice dumplings of Beijing differ from those in the south in that they are bigger in size. Fillings are mostly jujube or red bean paste, while some are stuffed with candy fruits.

Ingredients include glutinous rice, Chinese small iris, which is used to tie up the dumplings, reed leaves as wrappings, and jujube as the filling. The leaves should be cleaned and boiled before used, and it is advisable to keep the smooth side inward when wrapping. The glutinous rice has to be soaked overnight.

原　料

糯米500克,马莲草30根,箬叶40张,红枣30颗。

制作方法

1. 箬叶先洗干净再煮软,光面向内。
2. 取箬叶叠成中心为圆锥形的斗,放入泡好的糯米,放上2颗小枣,再放入糯米,将斗口包严,用马莲草捆紧放入凉水锅中,盖上锅盖,中火煮熟即可。食用时可蘸蜂蜜、白糖、糖桂花。

备注:包粽子的箬叶不能太小,包的过程中手要按压紧,不然糯米会漏掉。

五毒饼

五毒饼是北方端午节特有的节令食品。初夏时节正是毒物滋生活跃的时候，因此古人会食用"五毒饼"消病强身，祈求健康。早在清代时期的《燕京岁时记·端阳》中就有记载："每届端阳以前，府第朱门皆以粽子相馈饴，并副以樱桃、桑椹、荸荠、桃、杏及五毒饼、玫瑰饼等物。"有文字记载可考的五毒饼有两种：一种是用枣木模子磕出来，上吊炉烤熟，出炉后提浆上彩，表面上再抹一层油糖，点心上有凸凹的花纹；一种是翻毛酥皮饼，然后盖上鲜红的"五毒"形象的印子。下面做的是前一种。

Five-Poison Cake

A seasonal food for the Duanwu or Dragon Boat Festival on the fifth day of the fifth month of the Chinese lunar calendar, five-poison cake or Wudu Bing means driving away the evil spirits and praying for blessings. As some poisonous insects and animals are active in summer, ancient Chinese people developed the custom of eating cakes carved with images of five of the most typical of them, namely, scorpion, toad, spider, centipede and serpent, hence the name of the cake, five-poison cake.

Ingredients include red bean, brown sugar and vegetable oil for bean paste; lotus seed, sugar and salad oil for lotus seed paste; jujube, crystal sugar, sunflower seed oil and dry starch for jujube paste; fresh rose petal, sugar and salt for rose filling; shelled peanut, watermelon seed, walnut, pine nut and sesame, crystal sugar and lard for five-kernel filling, plus cooked wheat flour, syrup and salad oil.

原 料

A： 1.豆沙馅：红小豆500克，红糖250克，植物油适量。
　　2.莲蓉馅：莲子300克，白糖180克，色拉油150克。
　　3.枣泥馅：红枣500克，冰糖50克，清水50克，葵花油30克，澄粉30克。
　　4.玫瑰馅：鲜玫瑰花瓣500克，白糖1500克，盐20克。
　　5.五仁馅：花生仁、瓜子仁、核桃仁、松子仁、芝麻各100克，冰糖50克，白糖350克，猪油150克，青红丝50克，熟面粉100克。

B： 月饼面粉500克，糖浆750克，色拉油125克，水适量。

制作方法

1. 将A1的红小豆提前一天用凉水泡好，放入高压锅里，倒入清水没过红豆一指节多一点，盖上锅盖加阀，开大火煮红豆，上气后转中火高压30分钟，熟后水几乎就煮没了。准备一个盆，把漏网担在盆边沿上面，放进煮好的红豆，一边拿住漏网一边揉搓烂红豆，去除红豆皮。之后把豆沙馅放进锅里，上一勺油，将一部分红糖倒进锅里小火翻炒，直到把豆沙炒得软硬适中，豆沙起韧劲即可。

2. 莲蓉馅、枣泥馅做法同上。

3. 挑一些新鲜食用的玫瑰花600克，把玫瑰花瓣去除花萼花梗，得到约500克花瓣，加入500克白糖，装入广口瓶中盖盖密封，放在阴凉处，存放一年即可成玫瑰馅。

4. 将A5的果仁分别烤熟，将烤好的花生仁、核桃仁与冰糖分别擀碎，最后同瓜子仁、白糖、猪油、青红丝一起揉搓均匀即可。

5. 将B的原料混合成面团，揉透揉匀，揪成剂子待用。

6. 皮和馅比例为1：1。将馅一一包入皮肉，置于五毒饼模具内，放入200℃的烤箱烤熟即可。

备注：馅皮一定要包均匀，按入模具时也要注意力度的均匀，否则烤制出来的饼皮会薄厚不均。

糖火烧

糖火烧是北京著名的清真糕点，色泽棕红，吃在嘴里酥松绵软，香甜可口，不黏不腻，散发着浓浓的麻酱、红糖和桂花香气。

Sweetened Baked Wheaten Cake

A popular food for Beijingers' breakfast, sweetened baked wheaten cake, or Tang Huoshao, dates back 300 years.

Ingredients for the sweetened baked wheaten cake include wheat flour, yeast, brown sugar and sesame butter. The dough is kneaded in warm water and scalded for 30 minutes in the preparation.

原　　料

面粉500克，酵母5克，水35克，红糖100克，芝麻酱500克。

制作方法

1. 面粉加入酵母，倒入温水和成面团，烫30分钟。
2. 将红糖中的硬块擀细，与麻酱一起拌匀。
3. 将面团擀成长方形，均匀抹上麻酱、红糖卷起。
4. 卷好的卷略压扁，从两边向中间折叠。
5. 擀成长方形薄片，一切两半。
6. 取切好的一片，从两边1/4处向中间折叠两次，再擀成片，卷成卷。
7. 揪成一个个小剂子，两边收口向下捏紧成圆形。
8. 稍压扁，烤箱预热180℃烤约25分钟即可，凉食为宜。

备注：面粉要用标准粉，擀制时要快而稳，否则面皮会澥掉，包时要捏紧不露馅，否则烤时易糊。

羊眼包子

羊眼包子是回族饮食，相传清代康熙皇帝曾食用过羊眼包子，因其面皮暄软，羊肉馅细嫩多汁而非常喜欢，并由此而出名。因其个头小似羊眼，所以称作"羊眼包子"，包子虽小，但馅料鲜美，深受人们喜爱。

Sheep-eye Steamed Stuffed Bun

A Muslim food of Beijing, sheep-eye steamed stuffed bun, or Yangyan Baozi, got its name because of its size, which is about that of sheep's eye. Small as they are, they may have various fillings. It used to be appreciated by Emperor Kangxi.

Ingredients for typical sheep-eye steamed stuffed bun on the JE cuisine include gigot, winter bamboo shoot, scallion, ginger, salt, sugar, soy sauce, sesame oil, onion oil, mushroom, wheat flour and yeast.

原　　料

羊腿肉500克，冬笋100克，葱10克，姜10克，盐8克，糖6克，酱油20克，香油25克，香菇100克，面粉500克，酵母5克，水300克。

制作方法

1. 香菇用水泡好去根，切丁，冬笋切片。
2. 锅内加少许油，下葱、姜各10克，稍炒，待出香气时下香菇丁。冬笋片翻炒，待香菇炒出香气后即可出锅。
3. 将羊腿肉切成小拇指大小的方丁，下葱、姜、盐、糖、酱油，打上劲，再下香油拌匀，加入炒好的香菇丁和冬笋。
4. 将面粉、酵母、水一起和成面团，醒20分钟，再将其揉匀，揪成20克每个的剂子，擀开，包上肉馅，包成羊眼大小，醒发25分钟，蒸熟即可。

备注：羊肉选用内蒙的小尾寒羊，可保证肉馅不腥膻；面团要揉得均匀光滑为宜。

肉末烧饼

肉末烧饼乃宫廷御膳房的代表作之一。相传慈禧老佛爷梦见食用烧饼，醒来后用膳时，寿膳房恰巧正端上来肉末烧饼，竟和梦中的一模一样。老佛爷一高兴，立即赐名为"圆梦烧饼"。肉末烧饼外酥松、内暄软，肉香味浓。

Baked Sesame Seed Cake with Minced Meat

A representative of the royal cuisine of the Qing Dynasty, baked sesame seed cake with minced meat, or Roumo Shaobing, was what Empress Dowager Cixi wished to have in her dream, as the legend goes. Hence the cake is believed to stand for having one's dream come true. It is served with cooked minced meat sandwiched in the cake.

Ingredients include pork, water chestnut, chopped mushroom, winter bamboo shoot, minced ginger, straw mushroom, chopped scallion, chicken oil, soup stock, oyster sauce, wheat flour, baking powder, pepper powder, five spices powder and salad oil.

原　　料

猪肉500克，马蹄50克，香菇丁50克，冬笋50克，姜末40克，大葱花100克，鸡油40克，高汤150克，酱油100克。

A：面粉500克，发酵粉5克，水400克。

B：面粉300克，花椒粉10克，五香粉10克，色拉油300克，盐30克。

制作方法

1. 先将洗净切好的香菇丁、马蹄、冬笋丁汆水后干煸几下备用，另起一锅滑猪肉丁，放料酒、胡椒去腥味，炒至断生后沥干油分。
2. 再把锅烧热下鸡油，放入葱花、姜末爆香，加入煸好的香菇、冬笋、马蹄，炒香加入肉丁、高汤，下酱油、盐调味，收汁出锅备用。
3. 将原料A揉成面团、原料B混合拌匀待用。将300克色拉油加入锅中，烧至二成热后慢慢加入混合好的粉料中搅匀，放凉即可。
4. 将和好的面团擀成4毫米厚的长方形大片，将油酥匀抹在面片上，从上往下一层一层卷起，下成30克大小的剂子，将其由外往里收口，口对口收紧成一个球，光滑面沾上一层薄薄蛋液后蘸一层芝麻，放入200℃的电饼铛里，烙成两面金黄色出锅，夹上之前炒好的肉末即可。

备注：烙饼时先烙底部再烙正面，油稍多烧饼才酥松。

豆尔馒首

豆尔馒首属于带馅的馒头，豌豆馅香甜可口，馒首洁白暄软。

Steamed Bun Stuffed with Pea Paste

Called Dou'er Manshou, or steamed pea stuffed bread, steamed bun stuffed with pea paste was a regular snack served to emperors in the Qing Dynasty. The desirable pea is the sweet pea grown in Zhangjiakou on the north of Beijing.

Ingredients include cooked cake flour, pea, baking powder, sugar and yeast.

原　　料

低筋粉250克，豌豆150克，泡打粉2克，白糖20克，酵母2克，水适量。

制作方法

1. 豌豆泡软控干、蒸熟、压半碎，加白糖熬成豆沙馅。
2. 面粉加酵母、泡打粉和一起，成发面，压均匀，揪成每个15克的剂子备用。
3. 将发面包入豌豆馅，醒发后上笼蒸熟即可。

备注：熬制豌豆馅一定要沸腾才能去除豆腥味。

JE Mansion Cuisine *Desserts*

小肉粥

小肉粥也称鞑子粥，是满族的食品，后来传入宫中。将海参、冬菇、冬笋、猪瘦肉在一起熬煮，粥香浓郁，肉香和海参伴随在一起，食用后温胃驱寒。

Pork Porridge

A traditional food for rituals among the Manchu people, pork porridge, or Xiao Rouzhou, was favored by imperial families during the Qing Dynasty.

Ingredients include rice, lean pork, winter bamboo shoot, mushroom, sea cucumber and shredded scallion.

原　　料

大米50克，瘦肉20克，冬笋15克，冬菇20克，水发海参20克，鸡蛋清1个，淀粉30克，葱花3克，盐5克，胡椒粉5克。

制作方法

1. 将发好的冬菇、冬笋改刀成指甲片，猪瘦肉、海参改刀成比冬菇、冬笋稍大点的片，猪肉片加蛋清，淀粉上好浆待用。
2. 锅上火加入水，水和米的比例10:1，加入米熬到米开花约30分钟，粥稠浓时放入冬菇、冬笋、肉片、海参，加盐、胡椒粉调味，撒葱花即可。

备注：文火熬粥，米和水的比例一次加准确，不可中途加水。

健壹公馆菜·热菜

经典热菜分享
Hot Dishes

北京烤鸭

北京烤鸭历史悠久，早在南北朝的《食珍录》中已记有"炙鸭"。元朝天历年间的御医忽思慧所著《饮膳正要》中有"烧鸭子"的记载，烧鸭子就是"叉烧鸭"，是最早的一种烤鸭。而"北京烤鸭"则始于明朝。朱元璋建都于南京后，明宫御厨便取用南京肥厚多肉的湖鸭制作菜肴，为了增加鸭菜的风味，采用炭火烘烤，使鸭子口感酥香，肥而不腻。公元15世纪初，明代迁都北京，烤鸭技术也被带入北京并进一步发展。菜品选用北京京西玉泉山之水培育的北京填鸭为原料，经加工烤炙而成，皮酥肉嫩，呈枣红色，果木香气芬芳，为京师名馔。

Beijing Duck

Beijing duck, or Beijing Roast Duck, Beijing Kaoya, is the specialty of Beijing, or the most popular dish. It features crispy duck skin and tender meat, wrapped in thin pancake with scallion, shredded cucumber and sweet flour paste. Typical duck for the dish is cultivated by the water from Jade Spring in western Beijing.

Ingredients include crammed duck, scallion white, cucumber, sweet flour paste and thin pancake.

原　料

北京填鸭1只约2750克，葱白100克，黄瓜条100克，甜面酱75克，鸭饼20张，10比1饴糖水1000克。

制作方法

1. 从颈部向鸭体充气，再用小刀将右翅下开约3～5厘米长的月牙口，去除所有内脏，去翅尖和鸭掌。用水洗净内膛。用鸭掌支撑在脊骨和胸之间，以防胸脯塌陷，然后用鸭钩穿过胫骨挂起备用。
2. 烫胚：锅内放水烧开，一手提住鸭钩，另一手用开水淋浇鸭身约4次至均匀，趁热再淋浇饴糖水2次至均匀。
3. 晾胚：将烫过的鸭子挂在恒温恒湿的晾坯间，晾约1天待鸭皮见干即可。
4. 烤炙：在特制的烤鸭炉内，填足果木，待炉温升高至180℃、火焰温度达250℃时将晾好的鸭子挑入炉内鸭杠上，先烤右鸭身，后烤左鸭身，再烤鸭脊部，最后烤鸭脯，待鸭子上色至枣红时用杆挑起进行燎烤，使鸭全身颜色一致。烤约45分钟即成熟。再用鸭杆挑起鸭钩，使鸭背部向火，后手往后抽，前手左扭用力一拉，凭着惯性，将鸭身荡平，避过火苗，悠出炉门即可。
5. 片食：将烤好的鸭趁热放净膛水，片皮肉食用。片时要求片片带皮，片片都有肉。一般一只鸭子可片出90～120片之多。

备注：烤鸭去内脏时不要破坏它膛内尾部的油脂薄膜，否则易导致鸭坯变质。

灌汤黄鱼

灌汤黄鱼选料昂贵，鲍、参、翅、肚、虾肉、瑶柱、裙边全都囊括其中，是考验刀工、调味及火候的一道菜品。刀法纯熟才能使整条鱼脱骨且鱼皮完整；调味准确，鱼才能汤鲜味美；火候恰当，口感才能细腻滑嫩。

Yellow Croaker Stuffed with Hot Gravy

Large yellow croaker has been very expensive since ancient times and yellow croaker stuffed with hot gravy, or Guantang Huangyu, has been a challenge to even the most experienced chefs. After the fish is cleaned and gravy prepared with such expensive stuffs as swallow nest, shark fin and sea cucumber is put inside the fish belly, the fish would be steamed, boiled, steamed again and fried, with the process repeated time and again. In the process the shape of the fish must not be deformed. It would be considered a special treat even on the imperial dinner table.

Ingredients include large yellow croakr, shark fin, abalone, scallop, Amyda Sinensis carina, shelled prawn, mushroom and winter bamboo shoot.

原　　料

有机大黄鱼一条1500克，发好的鱼翅50克，发好的鲍鱼50克，瑶柱20克，发好的辽参50克，裙边20克，发好的鱼肚30克，大虾肉20克，虾胶50克，清鸡汤800克，发好的冬菇20克，冬笋20克，葱20克，姜20克，料酒30克。

制作方法

1. 将整条黄鱼打鳞去腮，可从鱼腮处用竹刀去骨和腮，下葱、姜、料酒腌制入味。
2. 将鱼翅先用鸡汤煨入味道，将鲍鱼、瑶柱、大虾肉、裙边、鱼肚、冬菇、冬笋全部切成细丝加汤加底味蒸30分钟后沥去汤。再加入汤，上锅烧开，调好味道用绿豆粉勾欠，制成鱼肚内的汤，从鱼腮部灌入鱼肚内，并用虾胶封口。
3. 将灌好汤的鱼放入油锅炸至表皮金黄，另取锅下底油，下入葱、姜、起锅下鸡汤，加入炸好的鱼烧15分钟收汁出锅即可。

备注：整个制作过程轻拿轻放，操作要娴熟，如果有一个环节出现了失误导致瑕疵，就宣告这道菜品失败。

酸菜鱼翅

宫廷菜代表菜品，原材料搭配合理。酸菜天然的酸味，使得菜品酸鲜适口，还可以帮助消化鱼翅，既利于吸收营养，又解酒开胃。

Shark Fin Soup with Pickled Cabbage

A representative dish of royal cuisine, this soup features a balanced combination of the main ingredient, shark fin, and the natural sour taste of pickled Chinese cabbage. The pickled cabbage can aid digestion, allowing the nutrition of the shark fin to be absorbed more efficiently, and this soup is an excellent starting dish.

Ingredients include shark fin, pickled Chinese cabbage, salt and meat broth.

原　　料

水发的鱼翅50克，酸菜30克，盐3克，白肉汤150克。

制作方法

1. 酸菜去除叶子选菜帮，切成2毫米细丝备用。
2. 将鱼翅用白肉汤煨制入味，加入酸菜再煨制3分钟，加盐调味即可入鱼翅盅内上桌。

备注：酸菜必须选天然发酵的，否则易出现酸涩味，而影响菜品鲜味。

健壹公馆菜・热菜

燕窝烩红白鸡丝

红鸡丝是熏鸡丝,白鸡丝是蒸熟的鸡胸肉丝,品尝时既有熏鸡丝的独特香气,又有白鸡丝细嫩,还有用清鸡汤煨好的燕窝的鲜味,菜品口味多变,层次分明。

Edible Bird's Nest with Braised Red and White Chicken

Red chicken is smoked chicken, and white is steamed chicken breast; along with the refreshing taste of edible bird's nest steeped in chicken broth, this dish features a unique layered combination of different flavors.

Ingredients include edible bird's nest, steamed chicken breast, smoked chicken breast, chicken broth, salt and starch.

原　　料

水发燕窝40克,蒸熟的鸡胸肉10克,熏鸡胸10克,清鸡汤150克,盐3克,水淀粉15克。

制作方法

1. 将熟鸡胸肉和熏鸡胸肉切细丝备用。
2. 锅中加入清鸡汤,再加入燕窝和切好的红白鸡丝,加盐调味烩制,然后淋水淀粉勾薄芡,装入燕窝盅内即可。

备注:勾芡时推芡,不可用手勺搅动,否则易将鸡丝搅碎,影响菜品美观度。

一品官燕

燕窝富含蛋白质，其中主要营养成分有精氨酸、胱氨酸、赖氨酸、糖类、钙、磷、钾等成分。具有生津润燥，养颜美容的功效。

Steamed Edible Bird's Nest

Edible bird's nests are rich in protein, as well as calcium, potassium and many other nutritious elements. This dish is invigorating and can aid in enhancing one's skin.

Ingredients include edible bird's nest, chicken broth and salt.

原　　料

水发燕窝50克，清鸡汤150克，盐3克。

制作方法

1. 将发好的燕窝择洗干净，加入清鸡汤，加盐调味盛入燕窝盅内，上蒸锅蒸10分钟即可。
2. 发好的燕窝也可配合果汁、蜂蜜、冰糖水制作成甜品。

备注：燕窝制作过程中避免接触碱性物品，因为在碱性环境下，燕窝的营养价值会降低。

水晶丸子

因其选用高档原料制作了一个水晶球酿在了肉馅中间,象征着有头有脑所以又名"全脑狮子头"。其难度在水晶球的软硬把握上,既要有脑的质感,又要它透明如水晶。据传慈禧老佛爷曾对此菜品的制作有过要求,说有一天寿膳房给老佛爷上了一道菜,报菜名曰红烧狮子头,老佛爷用筷子夹食时发现就是一个大肉丸子,就质问道:"既然曰头,为何无脑,难道你们也没有脑子吗?"这话马上传到了寿膳房,厨师们赶紧商量怎样才能有头有脑,后来就用上等食材制作出了水晶球酿在丸子中间,这样既符合了老佛爷追求极致享受的心理,又符合了菜品选料讲究的宫廷风范。菜品呈上后,老佛爷夹开看到水晶球晶莹剔透,各种高级食材随着惯性晃晃悠悠,灵动十足,吃一口,软嫩鲜香,顿时心花怒放,大加赞赏,这道菜品故而得以传承下来。

Crystal Meat Ball

A famous dish on the royal cuisine, crystal meat ball, or Shuijing Wanzi, features fillings made of ground pork belly with egg, salt and starch wrapping crystal balls made of ground water chestnut, shredded winter bamboo shoot, sea cucumber and mushroom shaped in egg shells. It tastes fresh and tender.

Ingredients include pork belly, shark fin, shredded winter bamboo shoot, shrimp, dried scallop, sea cucumber and mushroom.

原 料

五花肉150克,水发鱼翅80克,海参80克,香菇60克,虾肉60克,瑶柱60克,冬笋60克,盐10克,鸡蛋120克,淀粉100克,清鸡汤500克。

制作方法

1. 五花肉切成小粒,加盐、淀粉、鸡蛋,打成肉馅备用。
2. 将海参、香菇、冬笋和虾肉切细丝备用。
3. 将切丝的配料下锅焯水去除腥味,另起锅加汤调味,放入焯过水的配料,用绿豆粉勾芡至浓稠,灌入鸡蛋壳内晾凉定型。
4. 取肉馅将定型的水晶球包好下油锅炸至金黄,捞出下入调好味的汤中,小火炖煮2小时捞出装盘,将原汁收浓浇在丸子上即可。

备注:水晶丸子的馅料用汤一定要用清鸡汤,这样才可透明如水晶,又滋味鲜美。

乌龙戏珠带碗片

海参自古以来被奉为"八珍"之一，在宫廷菜的烹饪中尤其讲究。菜品选用辽东半岛刺参，成菜的辽参刺长挺拔，酱香棕亮，再配上鲜虾制作的丸子，象征着夜明珠，代表财富和地位。海参味道香浓，口感软糯。

Sea Cucumber with Shrimp Ball

The royal cuisine was fastidious about ingredients and the sea cucumber for this dish must be thorny sea cucumber from the Liaodong Peninsula, which stands for black dragon in the Chinese name of this dish, Wulong Xizhu. The shrimp ball stands of the pearl, and their combination suggests fortune and power.

Ingredients include thorny sea cucumber, shrimp, salt, soy sauce, sugar, cooking wine, soup stock and scallion.

原　　料

发好的40头辽参10条，虾肉100克，盐10克，酱油5克，糖8克，料酒15克，高汤200克，大葱250克，雪花粉500克。

制作方法

1. 将雪花粉和成面团，放置30分钟擀成面片制成汤饼。
2. 将虾肉制作成虾胶，再挤成丸子上锅蒸熟待用。
3. 锅上火加底油，下入大葱煸炒出香味，下入辽参在葱油中煎炒，烹入料酒、酱油、鸡汤，加盐调味，用糖色调色，烧入味后大火收汁装盘，旁边放上虾丸即可。汤饼煮好浇上葱烧汁装入小碗内，摆在盘子中即可。

备注：煸炒葱油时用中火煸炒，火太小会出熬葱味。

冰糖哈士膜

中国林蛙,是世界上唯一集药用、滋补保健和食用于一体的名贵经济蛙种,其雌蛙输卵管干燥品是名贵的中药材哈士膜,食用被誉为山珍。冰糖哈士膜口味甜香,入口即化,有养颜美容、滋阴强肾、抗疲劳、增加免疫力的功效。

Stewed Snow Frog with Crystal Sugar

Snow or forest frog is rather unique to northeast China's Heilongjiang, Jilin, and Liaoning provinces. Previously available only to emperors, the soup made of snow frogs has been a delicacy for centuries. Stewed snow frog with crystal sugar, or Bingtang Hasma, is believed both delicious and good for keeping one's youth.

Ingredients include soaked snow frog and crystal sugar.

原　　料

水发哈士膜150克,冰糖50克。

制作方法

1. 哈士膜用水加姜汁焯一下,去一下腥味沥干待用。
2. 哈士膜加入炖盅中,然后注入冰糖水,上火蒸25分钟即可。

备注:一定要去干净哈士膜的腥味。

健壹公館菜・热菜

万字扣肉

万字扣肉是清宫寿膳房为慈禧太后做寿时必用的菜品，软烂鲜香，突出食材本味并呈现出淡淡的雪菜的香气，肥而不腻。里边的鹿肉代表禄，五个装饰物代表着福、禄、寿、喜、财，五福捧寿。万字花刀代表着万寿无疆、万事如意、家庭幸福永久之意。

Svastika Braised Pork

A must for the longevity feasts for Empress Dowager Cixi since her 50th birthday, Svastika braised pork, or Wanzi Kourou, is fat but not greasy. While in the shape of Svastika, which symbolizes longevity, the decors stand for fortune, officialdom, happiness and wealth. The pork chunk must have eight layers each of fat and lean meat, with each lean meat layer having to be double the fat layer in thickness

Ingredients include refined pork belly, venison tenderloin, potherb mustard, scallion, aniseed, salt, sugar and soy sauce.

原　　料

精五花2500克，鹿后腿肉500克，雪菜500克，葱20克，八角2颗，盐2克，糖5克，酱油50克。

制作方法

1. 将五花肉入锅中水煮15分钟，再上锅蒸40分钟取出。在猪皮上抹上老抽，入油锅炸至上色后冷藏待用。
2. 将鹿肉切成肉末，加入酱油炒熟待用。雪菜切末后，加入酱油、清汤炒熟待用。
3. 将冷藏过的五花肉切成长、宽各4厘米的正方块，用万字刀法将其切成约1毫米左右的薄片，把切好的片圈好后放入碗中，上面盖上一层肉末和雪菜，然后加入葱段、大料、酱油和水，上锅蒸3个小时即可。

备注：只加葱和八角，这样可突出菜品所需的筵席味即甘鲜香的肉的本味。

JE Mansion Cuisine *Hot Dishes*

健壹公馆菜・热菜

油焖大虾

选用上等4头海虾制作。虾的营养价值极高，能增强人体的免疫力。海虾是可以为大脑提供营养的美味食品，油焖大虾色泽亮丽，鲜香适口，口感鲜嫩多汁。

Braised Prawn

A popular way to cook braised prawn, or Youmen Daxia, today is to use tomato sauce for the color. But that sauce is never allowed for this dish in the royal cuisine. Therefore the braised prawn cooked in royal culinary style keeps its original flavor while bright in color.

Ingredients include prawn of the size that four prawns weigh half a kilo, scallion, ginger, garlic, soy sauce, sugar, salt and shrimp sauce.

原　　料

4头大虾1只，葱、姜、蒜各10克，酱油2克，糖25克，盐3克，虾油10克。

制作方法

1. 将4头大虾，剪去虾脚，从虾背开刀，挑去虾线，剪去虾枪，去除沙包。将处理好的大虾入七成热的油锅里，炸至虾皮酥脆、沥油。
2. 另起锅放入虾油少许，下入葱、姜、蒜，同虾一起煸炒出香味，烹料酒和酱油，加虾汤、盐、糖等调味，加盖烧3分钟，大火收汁，汤汁变浓，滤去葱、姜、蒜末，出锅装盘即可。

备注：虾烧制的时间不宜过长，否则虾肉口感易老。

抓炒大虾

抓炒大虾也是"四大抓炒"之一,在慈禧年间,以御厨王玉山善做,汁明芡亮,大虾外酥里嫩,口味酸甜咸鲜,很受食客追捧。

Stir-Fried Prawn

Another of the four major stir-fried dishes on the royal cuisine of the Qing Dynasty, stir-fried prawn, or Zhuachao Daxia, is sweet and sour and appeals to patrons of all age groups.

Ingredients inclue prawns, sugar, vinegar, salt, soy sauce and wet starch.

原 料

虾肉400克,糖50克,醋30克,盐10克,老抽5克,湿淀粉150克。

制作方法

1. 将大虾肉开背取出沙线,用湿淀粉将大虾抓裹均匀,下油锅炸至外焦里嫩。出锅备用。
2. 锅下糖和醋调成小酸甜味,下淀粉勾芡,下入炸好的虾翻炒均匀出锅即可。

备注:将大虾下入油锅内,在220℃油温下要复炸一次,这样炸出的虾才外焦里嫩。

健壹公馆菜·热菜

御膳香糟鱼

据说明朝隆庆年间兵部尚书郭忠皋回乡探亲，从老家福山将一名厨师带回京都，适逢穆宗皇帝朱载垕为宠妃做寿，宴请文武百官，郭尚书便推荐福山厨师主持御宴。那厨师使出全身技艺，令御筵一扫旧颜，满朝文武无不开怀畅饮。朱载垕至翌日，日上三竿，方才酒醒，品之口中仍然美味不绝，对福山厨师深为叹服。数年后，那位厨师告老还乡。一日，朱载垕龙体欠安，不思饮食，甚念那位福山厨师做的"糟溜鱼片"，皇后娘娘派半副銮驾赶往福山降旨，将那名厨师和两名徒弟召进宫来。那名厨师的家乡被后人称为銮驾庄，就在今天的山东烟台福山区。

Mandarin Fish Fillet with Wine Sauce of Imperial Kitchen Style

Originally a cooking style in southern Jiangsu and Zhejiang, mandarin fish fillet with wine sauce of imperial kitchen style, Yushan Xiangzaoyu, is unique in using as a main ingredient distiller's grains left over from yellow wine making, which would be sealed in storage for at least six months. The wine sauce made of the distiller's grains give off a special aromatic flavor.

Ingredients include mandarin fish, wine sauce made of distiller's grains left over from yellow wine making, water chestnut, winter bamboo shoot, starch, sugar, salt and soup stock.

原　料

黑鱼1条，冬笋50克，马蹄50克，香糟100克，淀粉50克，白糖50克，盐15克，高汤15克。

制作方法

1. 黑鱼宰杀后取肉，头尾清洗干净，上锅蒸熟。
2. 鱼肉去刺骨后片成鱼片，上浆。
3. 炒锅放油烧热，把鱼片过油后备用。香糟、清汤放在一起淋芡粉，芡汁透亮时下入鱼片，轻轻翻匀，出锅放在鱼盘中间。
4. 再摆上头尾，此菜色泽米黄、清雅大方。
5. 糟卤制法：黄酒1000克、酒糟200克、冰糖50克、糖桂花30克混合在一起，装入纱布袋中放入冷藏室内吊糟。

备注：鱼片易碎，操作时动作要轻，以免鱼片破碎。

赛螃蟹

以黄花鱼为主料,配以鸡蛋,加入各种调料,炒制成的菜肴。黄花鱼肉雪白似蟹肉,咸蛋黄金黄如蟹黄。此菜鱼蛋软嫩滑爽,味鲜赛蟹肉,不是螃蟹,胜似蟹味,故名"赛螃蟹"。此菜色泽米黄,蘸着姜醋汁同食,的确是真假难辨。

Crab-Flavor Fish

With yellow croaker as the main ingredient, crab-flavor fish, or Saipangxie, tastes soft and tender. Served with ginger and vinegar, it often reminds people of the fresh crab meat.

Ingredients include deboned yellow croaker, salty egg yolk, peeled ginger chops, egg white, wet starch, rice vinegar, and water that has soaked dry shelled shrimps.

原　　料

净黄鱼肉150克,咸蛋黄30克,姜米20克,蛋清10克,湿淀粉20克,米醋20克,海米水50克。

制作方法

1. 将黄花鱼去骨,然后去皮,剔出净鱼肉,改成1厘米的菱形小方丁,上浆待用。
2. 将油锅上火,下油烧热倒入浆好的鱼肉划开,倒出锅肉留底油,下咸蛋黄炒成小块状,下蛋清,咸蛋黄外蒙一层蛋清,熟后盛出。
3. 另起锅下汤,加海米水、盐、胡椒粉,调味下入后,下入鱼肉淋芡,淋上姜醋汁出锅装盘,撒上香菜末即可上桌。

备注:咸蛋黄有咸味,用盐调味不可过重。

健壹公馆菜·**热菜**

烧羊肉

烧羊肉是北京清真名肴，外焦里嫩，入口不腻，色泽金黄，最宜就烧饼和佐酒食。

Braised Lamb

Braised lamb, or Shaoyangrou, is a Muslim delicacy of Beijing. Crispy outside but tender inside, it is fat but not greasy and good to eat with baked sesame cake and liquor.

Ingredients include lamb breast, scallion, ginger, garlic, cucumber, lotus-leaf-like pancake, coriander herb, sweet fermented flour sauce, pepper salt, aniseed, bark of Japanese Cinnamon, salad oil, sesame oil, pricklyash peel and white sesame.

原　　料

羊腩肉1500克，葱100克，姜100克，黄瓜50克，荷叶饼30张，甜面酱100克，花椒盐10克，大料10克，桂皮5克，香菜15克，色拉油1000克，香油30克，花椒10克，白芝麻25克。

制作方法

1. 精选羊腩肉焯水。用葱、姜炒香加料酒和水等把羊腩炖熟烂，2小时左右后出锅。
2. 炖好后将羊肉捞出放凉，汤过滤装入容器。
3. 将白芝麻用锅炒香，跟花椒盐一起倒入小碗做蘸料用。
4. 炒锅中倒入香油，烧至六成热后放入羊肉块。炸至两面焦黄，捞出沥油切成条装盘，撒上白芝麻点缀香菜，蘸白芝麻、花椒盐食用。也可配饼、葱丝和瓜条食用。

备注：炸制时一定要炸至焦黄，这样才不腻。

健壹公館菜·热菜

烤哈尔巴子

"哈尔巴子"系满语,汉译即肩肿骨,菜品选用肩胛骨附近的肉,因为筋肉相间,吃起来会口感柔韧,咀嚼感非常好,同时肉香浓郁。

Roast Harba

In Manchu language, "harba" means shoulder blade. Roast Harba, or roast pork shoulder, is a traditional Manchu delicacy and was high on the royal menu of the Qing Dynasty.

Ingredients include pork shoulder blags, scallion, cooking wine, soy bean paste and light soy sauce.

原 料

哈尔巴子500克、葱1根、香油1000克、料酒30克、黄酱30克、生抽20克。

制作方法

1. 将哈尔巴子肉加葱、姜、料酒、黄酱、生抽腌制入味。
2. 将腌好的哈尔巴子放入烤炉烤熟备用。
3. 上菜时用香油烧热,将烤好的哈尔巴子炸至外酥里嫩,改刀上桌即可。

备注:烤制时不可急火快烤,要文火保证肉外酥里嫩。

健壹公馆菜·热菜

鸡米鹿筋

鸡米鹿筋看似平淡，其实用功不少，看似淡而无味，其实吃一口滋味醇香，这就是中国菜内敛含蓄、低调的表现。口感柔韧，口味鲜香醇厚。

Diced Chicken with Venison Tendon

Ingredients for diced chicken with venison tendon, or Jimi Lujin, include dry venison tendon, chicken tenderloin, refined salt, pepper powder, winter bamboo shoot, wet starch, egg white, yellow rice wine and scallion.

原　料

干鹿筋250克，鸡里脊50克，精盐2克，胡椒粉0.5克，冬笋50克，淀粉15克，鸡蛋清2个，黄酒5克，葱10克，清鸡汤500克，鸡油50克，花生油1500克。

制作方法

1. 用火燎去干鹿筋上的鹿毛，洗净后，剁成长16厘米的段。将炒锅置于火上，倒入花生油烧到三四成热，放入鹿筋约炸15分钟，炸到鹿筋弯曲、起泡并透明时，捞出沥油。将炸过的鹿筋放入热水盆里，泡1小时后洗去油，将水滗出，然后用开水氽10分钟取出。
2. 将鸡里脊肉剔去筋，切成小方粒，放入碗中，加鸡蛋清、盐搅拌，冬笋切成长5厘米、宽1.6厘米的片，葱切成葱米。
3. 将炒锅置火上，加入鸡油烧热，下葱米、姜末，随即放入鸡米炒香，加入鹿筋、清鸡汤、盐、黄酒、胡椒粉及冬笋片，烧开后，再烧3分钟入味收汁，淋上鸡油即成。

备注：鹿筋的处理一定去掉它本身的腥膻味，另外是清鸡汤，这道菜品的鲜味完全依靠汤的鲜美。

椒盐肘子

肘子外焦里嫩,肥而不腻,配荷叶饼和葱丝、椒盐食用,咸香可口,回味悠长。

Pork Shoulder with Pepper and Salt

Crispy outside and tender inside, pork shoulder with pepper and salt, or Jiaoyan Zhouzi, is also served with lotus-leaf like thin pancake.

Ingredients include deboned pork shoulder, salt, scallion, ginger and pepper.

原　　料

猪后肘1个,盐15克,葱20克,姜20克,椒盐10克。

制作方法

1. 将肘子上火烧净外边的毛,刮洗干净,下锅焯水,然后剔掉骨头,抹上酱油,炸至枣红,放入调好的汤中,上锅蒸3小时至肉软烂出锅。
2. 将蒸好的猪肘放在八成热的油锅里炸至肉皮起泡焦黄,改刀配上春饼即可。

备注:肘子在炸制时一定要炸得快焦糊时才可以,焦而不糊的火候才可以保证肘子肥而不腻,咸香可口。

酸菜炉肉锅子

炉肉味道独特，烧烤的肉香气浓郁，再配酸菜解腻又开胃，如果加入酱豆腐、韭菜花、辣椒油，风味又更加多变而独特。

Pork Belly Cooked with Pickled Cabbage

Ingredients include pork belly and pickled cabbage.

原　　料

五花肉750克，酸菜300克，葱20克，姜20克，盐15克，辣椒油30克，香菜末30克，韭菜花30克、酱豆腐30克。

制作方法

1. 五花肉拿松肉针插眼，加盐、葱、姜搓匀，加料酒放入容器内腌制入味后取出，刮净上面的杂物，下开水锅烫皮，晾干。
2. 进烤鸭炉烤至肉皮起泡，取出。再加汤上蒸锅蒸1小时后晾凉。
3. 把晾凉的五花肉切成3毫米片，加葱、姜放入碗中，加高汤蒸20分钟后取出备用。
4. 另起锅加汤，下酸菜和蒸好的炉肉，打去浮沫，调味，配4个料碟（辣椒油、香菜末、韭菜花、酱豆腐），即可上桌。

备注：此菜制作过程中，要有足够的时间，时间短，不易入味，影响其风味。

干炸丸子

干炸丸子是一道烧碟菜,讲究的是外表焦脆酥香,内肉鲜嫩。炸完后表层还能看到丝抱丝的感觉。

Dry-Fried Meatballs

Dry-fried meatballs have a crisp outer layer and tender meat filling. They are made by first deep-frying, then roasting on a grill.

Ingredients include pork, bean curd sauce, starch, salt, spiced salt and onion and ginger water.

原　　料

肥瘦肉馅300克,黄酱5克,淀粉20克,盐3克,椒盐15克,葱姜水15克。

制作方法

1. 将剁好的肥瘦肉馅加入黄酱、淀粉、盐和葱姜水,搅拌均匀至上劲,下油锅炸制外焦里嫩出锅装盘即可。
2. 上桌配花椒盐食用,也可蘸老虎酱或苜蓿肉,一道干炸丸子可以三吃。

备注:炸丸子时要把炸过的丸子用漏勺捞出油面,上下颠动以降低表层温度再复炸,反复几次,丸子才可外焦里嫩。

炸鹿尾

炸鹿尾不是鹿尾胜似鹿尾，有鹿尾淡淡的野味，有松子的坚果香气，再配上蒜汁辛辣鲜香，风味别致。

Deep-Fried Deer's Tail

Called deep-fried deer's tail, or Zhaluwei, the dish has nothing to do with the deer. All involved is but pork from pig haunch, which is ground with shelled pine nut, egg white and grounded nutmeg as fillings to be filled into sausage casing, marinated and fried. The fried pork case looks like deer's tail, hence the name.

Ingredients include pork from pig haunch, pig liver, sausage casing, shelled pine nut, egg white, amomun fruit powder, grounded nutmeg, pepper powder, ginger root extract, scallion, sugar, salt, soy sauce and starch.

原　　料

猪前尖肉500克，猪肝100克，猪肠衣1根，松仁50克，蛋清2个，沙仁粉5克，白豆蔻粉10克，盐10克，葱汁50克，姜汁50克，花椒粉5克，淀粉100克，盐13克，糖3克。

制作方法

1. 把前尖肉和猪肝剁碎待用。
2. 把松仁炒出香味加入到剁好的馅中，加蛋清、盐、沙仁粉、白豆蔻粉、花椒粉、葱姜汁拌匀，再加入淀粉，打上劲后灌入肠衣，卤制30分钟。
3. 卤汁的配制：猪棒骨1500克，用水煮开锅打去浮味，放入铁桶中熬成高汤，加入老抽200克、生抽100克，用料包布将沙仁100克、豆蔻100克包好放入汤中，加葱、姜，入桶小火煮制1小时即可。

备注：灌制肠衣时不能灌得太满，煮时肠衣要扎眼，这样不易爆裂。

干炸佛手

北京名菜"炸佛手"最早为清宫菜。传说,有一位仙女把自己的双手抛向人间,化成佛手柑,给一位英俊的青年猎人治好了病。佛手柑,色鲜美、香气浓郁,一直深得人们喜爱,清宫御膳房的厨师便照佛手柑的形象,制作了"炸佛手"这道名菜。菜品咸鲜酥香,回味无穷。

Dry Fried Citron Roll

Another dish of the royal cuisine in the Qing Dynasty, dry fried citron roll, or Ganzha Foshou, got the name as the roll is in the shape of finger citron.

Ingredients include ground pork filling, eggs, salt, ginger root extract, pepper powder and starc.

原料

猪肉馅300克,鸡蛋2个,盐3克,姜汁5克,胡椒粉3克,淀粉30克,花椒盐适量。

制作方法

1. 将猪肉馅加淀粉、姜汁、盐搅拌均匀备用。
2. 将鸡蛋打开,上锅摊成蛋皮。
3. 将摊好的蛋皮,抹上搅拌好的肉馅卷成长条,放入冰箱冷冻成型后,取出改刀成佛手,下油锅炸至金黄酥脆,即可装盘配椒盐上桌。

备注:炸佛手的时间不可过长,否则容易炸干。

抓炒里脊

"四大抓炒"之一的抓炒里脊，口味小酸甜，口感外酥里嫩，风味独特。

Stir-Fried Tenderloin

One of the four major stir-fried dishes on the royal cuisine of the Qing Dynasty, stir-fried pork tenderloin, Zhuachao Liji, has been regarded as representative of old Beijing delicacies.

Ingredients include pork tenderloin, fresh ginger, sugar, vinegar, salt, cooking wine, soy sauce and starch.

原　　料

猪里脊200克，鲜姜50克，糖20克，醋15克，盐3克，料酒10克，酱油10克，淀粉100克。

制作方法

1. 鲜姜打成姜汁，把糖、醋、盐、酱油兑成碗汁备用。
2. 里脊肉切成柳叶片，加淀粉抓匀后入锅，炸至金黄色。
3. 净锅上火，把兑好的碗汁倒入锅中，下入里脊肉翻炒均匀，全部裹上味后，淋明油出锅即可。

备注：处理里脊肉时，一定要先除去连在肉上的筋膜，否则不但切不好，吃起来口感也不佳。

抓炒腰花

抓炒腰花是"四大抓炒"之一，要求麦穗花刀清楚美观，口味酸甜鲜香，汁明芡亮，外焦里嫩。

Sauteed Pork Kidney

This is one of the four main sautéed dishes in royal cuisine, and features distinct and complex knife work, a robust flavor, and a tender and juicy texture.

Ingredients include pork kidney, white sugar, rice vinegar, soy sauce, salt, onion and ginger water and starch.

原　　料

鲜猪腰200克，白糖30克，米醋20克，酱油10克，盐5克，葱姜水10克，淀粉80克。

制作方法

1. 鲜猪腰用平刀法片开，去腰骚和筋膜，改刀成麦穗花刀，用葱姜水腌5分钟，再用干毛巾吸干多余水分备用。
2. 用小碗将米醋、酱油、白糖、水、盐和淀粉调成芡备用。
3. 油锅烧热，腰花拍干淀粉，下入油锅炸至外焦里嫩时出锅。锅中留少许底油加入腰花，烹入碗芡快速翻炒均匀出锅装盘即可。

备注：抓炒要体现的是手眼配合，快速成菜。如果操作慢，腰花就会口感变老，失去此菜的风味特点。

JE Mansion Cuisine *Hot Dishes*

健壹公馆菜·热菜

焦熘丸子

焦熘丸子是鲁菜中很有特色的代表菜之一。其特色是色泽枣红、酥脆鲜香。

Sauteed Fried Meat Ball

Coke crispy outside, sautéed fried meat ball, or Jiaoliu Wanzi, is a representative of Lucai or Shandong cuisine, with a long standing history. The meat balls are crispy and taste great.

Ingredients include pork belly, fungus, sliced bamboo shoot, edible rape, salt, soy sauce, cooking wine, vinegar and starch.

原　料

猪五花肉300克，木耳20克，笋片20克，油菜心20克，盐5克，酱油5克，料酒5克，香油3克，淀粉10克，色拉油500克。

制作方法

1. 将五花肉去皮，剁成肉馅，加淀粉打成肉馅。
2. 炒锅上火，加色拉油烧至四成热，将肉馅挤成丸子下入锅中，反复炸至外酥里嫩，即可出锅。
3. 用盐、酱油、料酒、淀粉调成碗芡，倒入锅中，将丸子与配料下锅拌炒均匀出锅即可。

备注：为保证丸子的焦香感觉，丸子要下锅复炸一遍。

酱爆熏鸡丝

酱爆熏鸡丝是道传统菜品,但是做法上有别于其它酱爆菜品,是用酱把鸡丝拌匀并过油后再炒,比传统酱爆的酱香味又多了一份干香。

Quick-Fried Shredded Smoked Chicken in Bean Sauce

The smoked chicken shall be debone and shredded it into thin strips for quick-fried shredded smoked chicken in bean sauce, or Jiangbao Xunjisi. .

IIngredients include smoked chicken, green garlic and soybean paste.

原　　料

熏鸡一只,青蒜50克,黄酱10克,淀粉30克。

制作方法

1. 将熏鸡折骨,把肉拆成细丝,备用。
2. 把鸡丝用黄酱拌匀,再撒上白面拌匀。
3. 锅烧油至五成热,将拌好的鸡丝下入油锅炸至定型,出锅。
4. 锅留底油,加青蒜丝,下入炸好的鸡丝,翻炒均匀,即可上桌。

备注:拆鸡丝时注意粗细均匀,以利于菜品炸制时控制火候。

炸豆腐角

豆腐因其营养丰富，是素菜品常用的食材，炸豆腐角是"素四供"之一，蘸着椒盐吃，豆香味浓郁。

Fried Bean Curd Puff

Boreal bean curd of hard texture is used for fried bean curd puff, or Zhadoufujiao, and the bean curd should be cut angularly into triangular slices of 3mm thick each.

Ingredients include bean curd and spiced salt.

原　　料

北豆腐250克，花生油800克，椒盐15克。

制作方法

1. 把北豆腐斜刀切成8毫米的三角片。
2. 锅烧油至六成热，将切好的豆腐角一片一片下入锅中，炸至金黄色即可出锅，上桌时配椒盐即可。

备注：炸的火候要够，豆腐外边要有一层酥皮，吃着才有外干香内鲜嫩的味道。

素丸子

素丸子属于"素四供"之一。选材讲究,外酥内嫩,蔬菜香味浓郁,营养丰富。

Minced Vegetables Bal

Crispy outside but tender inside, minced vegetables ball, or *Suwanzi*, is a dish unique to the royal cuisine of the Qing Dynasty. As it is meat free, the dish is picky about ingredients.

Ingredients include carrot, coriander herb stalk, gluten, mushroom, soft creak or pancake wrapping, pickled cucumber, bean curd, winter bamboo shoot, salt, pepper powder, chicken powder, sesame oil and starch.

原　　料

胡萝卜50克,香菜梗50克,面筋80克,香菇50克,软咯吱120克,酱瓜10克,豆腐75克,冬笋50克,盐3克,花椒粉5克,香油10克,淀粉60克。

制作方法

1. 把胡萝卜、香菜梗等切末挤一下水分。
2. 加入淀粉、盐、花椒粉、香油打成馅,揉成丸子下锅炸熟即可。

备注:软咯吱主要是保证丸子成型用的,注意控制好用量,过多则使丸子过软,不易成形。

咯吱盒

咯吱盒是过去炸货铺子售卖的食品，口感酥香，蔬菜味道丰富，吃完后口齿留香。

Fried Bean Flour Roll

A dish said to be older than the history of Beijing, fried bean flour roll, or Gezhihe, originated from the Shandong pancake, which is extremely thin. People love to wrap scallion with bean paste in it and wrap it up for a meal. But as pancake gets dampened in days it becomes softened and does not taste good. As the legend goes, some careful people rolled the pancake up, cut the roll into pieces and fried them in oil, which would be crispy and last longer. As it creaks at the bite, it got its popular name as fried creak. It has become a popular dish typical of Beijing.

Ingredients include soft creak or pancake wrapping, carrot, coriander herb, salt, pepper powder, starch, sesame oil and hard creak or pancakngwrapping.

原　　料

软咯吱500克，胡萝卜400克，香菜100克，盐6克，花椒粉10克、淀粉30克、香油1000克，硬咯吱皮150克。

制作方法

1. 将各种原料整理干净，胡萝卜刨丝，香菜切末备用。
2. 将软咯吱用手搓成泥状，加胡萝卜、香菜、花椒粉拌匀，加香油、盐拌匀。
3. 硬咯吱皮上用抹刀将咯吱馅抹匀，约5毫米厚，再将另一张硬咯吱皮铺在上面。用刀稍压改成6厘米长、2.5厘米宽的方块。
4. 将油锅上火烧到170℃，下咯吱盒炸4分钟，即可上桌。

备注：用刨刀刨的胡萝卜比切出来的香气更浓郁。

健壹公馆菜·热菜

栗子扒白菜

北京的一道传统风味菜肴。"白菜"的寓意为百财,而"栗子"是多子多福和吉利的象征。白菜软嫩鲜香,栗子香甜糯软。

Braised Chinese Cabbage with Chestnut

A traditional dish of Beijing, braised Chinese cabbage with chestnut, or *Lizi Pa Baicai*, stands for clean personality with fortune. Chinese folk customs hold chestnut as auspicious as it pronounces the same as "benefiting offspring."

Ingredients include shelled chestnut, thick chicken soup, ginger, sugar, salt, starch and chicken oil.

原　　料

白菜心300克,去皮板栗100克,浓鸡汤400克,姜汁10克,糖15克,盐5克,淀粉50克,鸡油5克。

制作方法

1. 板栗用鸡汤蒸熟放在一边待用。
2. 大白菜取白菜心,加鸡汤放入锅中烧,再加入蒸好的板栗、盐、糖、姜汁,白菜烧入味后收汁,淋少许芡汁,淋鸡油出锅扒在盘中即可。

备注:制作这道菜一定要选用新鲜板栗,因为它颜色金黄,口味甜糯。

拔丝西瓜

西瓜乃夏天消暑之佳品，清甜可口、止渴生津、人人喜爱。用西瓜烹制的拔丝西瓜球，银丝缭绕、外酥里鲜、香甜可口。

Candied Watermelon

Watermelon is cool as a summer fruit. Candied watermelon balls, or Basi Xiguaqiu, is one of the hot dishes made out of this cool fruit.

Ingredients include watermelon, wheat flour, sesame, sugar and baking powder.

原　　料

西瓜200克，淀粉250克，芝麻10克，白糖100克，泡打粉20克。

制作方法

1. 将西瓜去皮，切成1厘米见方的小块备用。
2. 将淀粉加泡打粉、水、和成脆皮糊。
3. 锅烧油至五成热，将切成的西瓜块挂上调好的脆皮糊下入油锅，炸至金黄，捞出沥油。另起锅，下白糖、加油、水，炒化至起米黄色小泡后，倒入炸好的西瓜块，翻裹均匀，撒上芝麻出锅装盘。上菜时配一小碗凉白开水，蘸水后方可食用，否则容易烫伤。

备注：在熬制糖液的时候要注意火候的拿捏，火候不够会反沙不出丝，过了则出苦味。

烩乌鱼蛋

乌鱼蛋是由雌乌鱼的卵腺加工制成的,含有大量的优质蛋白。乌鱼蛋饱满坚实、体表光洁,蛋层揭片完整,乳白色为上品。烩乌鱼蛋汤鲜美,微酸,微辣,口味柔和,又不失层次分明,造型朵朵如花。

Cuttlefish Roe Soup

Cuttlefish roe was one of the tributes to the imperial court up to the end of the Qing Dynasty. Cuttlefish roe soup, Hui Wuyudan, was one of the favorites on the royal menus. It should taste fresh, a bit salty and mildly hot.

Ingredients include cuttlefish roe, sesame oil, minced coriander herb, clear soup, ginger root extract, pepper powder and rice vinegar.

原 料

乌鱼蛋片100克,香油5克,香菜末15克,清汤500克,姜汁10克,胡椒粉15克,米醋15克,盐5克。

制作方法

1. 乌鱼蛋片焯水去除腥味和咸味后备用。
2. 锅中加入清汤,放入乌鱼蛋片,烧开锅后,加姜汁、胡椒粉、米醋、盐调好味,勾芡出锅即可。
3. 上面撒上香菜末,淋少许香油。

备注:乌鱼蛋有海水的咸涩味,一定要去除彻底。

健壹公馆菜·冷菜

精选冷菜分享
Cold Dishes

杏干肉

杏干天然的酸甜味道，再配上脆甜的莲藕，口感层次丰富，回味持久。

Dried Apricot Meat

Sour and sweet, dried apricot meat, Xinggan Rou, is appetizing. Not many restaurants bother to cook the dish with dried apricot, but just cut the meat in the size of it. Yet the JE Mansion takes pride in having dried apricot in it, making the dish taste great.

Ingredients include dried apricot after it is soaked in cold water, pork tenderloin, egg, peanut oil, corn starch, salt, crystal sugar and rice vinegar.

原　　料

猪小里脊肉500克，杏干100克，莲藕150克，糖桂花20克，鸡蛋1个，花生油500克，玉米淀粉20克，盐5克，冰糖50克，白醋50克。

制作方法

1. 杏干用开水泡两小时。
2. 将猪小里脊肉切成3毫米厚的片，加盐、鸡蛋、玉米淀粉搅拌均匀。
3. 将锅加热倒入花生油，烧至五成热，倒入拌好的猪肉片，炸至外酥里嫩捞出。莲藕去皮切片，焯至刚熟有脆感备用。
4. 将锅中加杏干水、冰糖、白醋，小火烧开，撇去表面的沫，将炸好的肉片倒入，烧3分钟后，再放入泡好的杏干，改中火慢慢收汁，将汁收到浓稠度能挂住肉片即可，盘底放上藕片，将杏干肉码放在藕片上，淋上糖桂花汁即可。

备注：选杏干时别选用杏脯，因为杏脯用糖腌渍过，没有杏干天然的酸甜味道。

蟹肉卷

选用肥美的梭子蟹肉,制作成蟹肉卷,美观大方,方便食用。菜品口味鲜甜,汁多肉嫩。

Crab Meat Roll

It doesn't matter whether it is meat inside egg wrapping or vice versa, as it must be attractive at dinner table. Those who are not sure of making fried egg wrapping well may add some wet starch or flour to the egg paste.

Ingredients include portunid, eggs, ginger, salt and wet starch.

原　　料

梭子蟹2只,鸡蛋2个,姜丝20克,盐5克,湿玉米淀粉10克。

制作方法

1. 梭子蟹上笼蒸8分钟,自然晾凉,用剪刀从腿部两边取出蟹肉,再去除蟹肉里面的软骨,备用。
2. 将鸡蛋打入容器中,加盐、湿淀粉搅匀,备用。
3. 将方形不粘锅烧至50℃热,上面用毛刷刷一层清油,将调好的蛋液分两次倒入锅内,做成蛋皮。
4. 将做好的蛋皮和拆下的蟹肉加姜丝卷起,最后斜刀切成段,装盘成菊花形即可。

备注:蟹一定要选择鲜活的,鸡蛋皮不能卷得太厚,否则会影响蟹肉的口味。

JE Mansion Cuisine *Cold Dishes*

御膳水晶虾

菜品透明如水晶,是厨师制作用心的体现,非常适合夏季食用。给食客的感觉就是清清凉凉,口味滑嫩,鲜味十足。

Sautéed Shrimp Aspic

Actually cooked pork skin aspic, sautéed shrimp aspic, or Shuijingxia, is a blended jelly of pork skin and shelled pink shrimps, which represents a variation of flavor and smart culinary art.

Ingredients include pink shrimps, raw pork skin, coriander herb leaf, rice vinegar, light soy sauce, minced garlic and sesame oil.

原 料

斑节虾150克，猪皮350克，香菜叶10片，碗汁（米醋10克、生抽10克、蒜泥5克、香油两滴），老鸡750克。

制作方法

1. 斑节虾去头去壳，并挑去虾线，洗净后吸干水分备用。
2. 老鸡打理干净，生猪皮放开水中烫5分钟，浸冷水后用刀刮净油脂，和老鸡一起加水，以小火炖成清鸡汤，撇干净油。
3. 将虾肉放入烧开的清鸡汤中加盐调味，再把它分别倒入模具中，每个里面放一个虾、一片香菜叶，进冰箱冷却成形。上桌前浇碗汁即可。

备注：清鸡汤的清澈度决定水晶冻的透明度。

陈皮兔肉

高蛋白、低脂肪的兔肉，既有营养，又不会令人发胖，是理想的美容食品，颇受年轻女子的青睐。陈皮兔肉，麻辣鲜香，口感细腻，陈皮香味突出。

Rabbit Meat with Tangerine Flavor

Using dried tangerine peel to flavor the dish, rabbit meat with tangerine flavor, Chenpi Turou, blends the fine meat of rabbit with the fragrance of tangerine peel, and is delicious with spicy ingredients.

Ingredients include chopped rabbit meat, dried tangerine peel, dried pepper, Chinese prickly ash, scallion, ginger, salt, sugar, yellow wine, peanut oil, sesame oil and pepper oil.

原　　料

兔肉约1000克，陈皮20克，花椒5克，干辣椒5克，葱段10克，姜10克，盐13克，糖3克，黄酒10克，花生油500克，香油10克，红油20克。

制作方法

1. 兔肉洗净，切块，用适量盐、黄酒、葱姜腌制入味，陈皮用适量开水泡出味备用。
2. 倒花生油入锅，烧至六七成热时下兔肉，微炸一下，稍上色即可捞出。
3. 锅里留少许油，下花椒、干辣椒、陈皮、葱、姜，炒香后加陈皮水烧开，再下炸好的兔丁，加盐、糖、黄酒，烧至兔肉入味后收汁，加香油、红油，翻炒均匀出锅即可。

备注：选用陈皮时一定要深褐色的陈皮，这样的陈皮，才可使菜品风味明显。

JE Mansion Cuisine *Cold Dishes*

健壹公館菜·冷菜

罗汉肚

罗汉肚是在猪肚内装入猪前肘肉蒸制而成，口味咸鲜，口感柔韧，乃佐酒好菜。

Pork Tripe Stuffed with Meat

Pork tripe stuffed with meat is known as "Arhat's Belly" or Luohandu in China, because the round shape and clear layers of the meat look like an arhat's belly. It uses pork tripe as a wrapping, and the pork tripe is often from piglet up to one year old.

Ingredients include fresh pork triggs, carrot, ginger, scallion whiin, fresh pork shouldil, corn starch and salt.

原　　料

新鲜猪肚1个约600克，新鲜前肘子300克，胡萝卜1根约40克，生姜20克，大葱白20克，盐13克，生粉100克。

制作方法

1. 将鲜猪肚用生粉揉抓10分钟，用剪刀将肚子里面的白色油脂去净，用清水洗净备用。
2. 将鲜肘子打理干净，把骨头剔出，将肘子肉切丁备用。
3. 将生姜、大葱白切好备用。
4. 用不锈钢盆将切好的肘子肉丁、姜末、大葱末搅在一起，顺一个方向搅制上劲，盐分三次放入，用手搅10分钟，拌匀，放入猪肚中，用鹅尾针封住肚口备用。
5. 将装好的猪肚，腌24小时入味，腌好的猪肚上笼蒸120分钟。
6. 猪肚蒸好后，用两个长托盘，将猪肚用重物压平，放凉，切薄片装盘即可。

备注：上笼蒸的前30分钟，每5分钟用竹签扎孔放掉里面的空气，防止蒸爆。

芥末墩

大白菜是过去北京老百姓冬季的看家菜，快入冬时都会储存，这样的白菜因为水分有所挥发，所以愈发显得清甜，用它做的芥末墩味道更是辛辣通气，清甜爽口。

Mustard Mound

A must for the lunar New Year dishes, mustard mound, or Jiemo Dunr, which is Chinese cabbage pickled in mustard, used to be a favorite of Manchu people. It is now a typical Beijing cold dish and is particularly cool and refreshing to go with oily food. The mustard juice is usually fermented two days in advance to get rid of its bitterness.

Ingredients include Chinese cabbage, yellow mustard powder, sesame oil, salt, rice vinegar and sugar.

原　　料

大白菜1棵，黄芥末粉200克，香油5克，盐6克，白醋50克，热开水100克，白砂糖15克，马连草15根。

制作方法

1. 将黄芥末粉加上白醋、热开水、白砂糖，全部倒入容器中搅开，盖上盖，发酵24小时备用。之后做成芥末糊。将马连草泡软备用。
2. 将白菜去老帮叶，切去根部，把中段切成2厘米高、直径3厘米的圆柱形，用马连草捆绑入开水锅氽烫至半熟，即捞出抹上芥末糊，一层一层码放在坛子内密封3天，发酵至芥末味道和白菜融为一体、香气扑鼻时开坛，取出加醋、香油即可食用。

备注：白菜不要煮太久，煮久了会影响它的口感。

榅桲白菜

榅桲和白菜拌制而成，有榅桲的酸甜味，加上白菜的清爽口感，颜色红白相间，是一道开胃的菜品。适合秋冬食用，润燥健脾。

Chinese Cabbage with Wenpo

Wenpo is the Chinese name of wild hawthorn, and its fruits are very small but very fragrant. It grows in the mountains in Beijing's northern suburb and ripens in autumn. That is why Wenpo Baicai, or Chinese cabbage with wild hawthorn, is a dish for winter, and it is unique to Beijing. The Chinese cabbage used for this dish must be the yellow heart of the vegetable, which is tender and sweet. The cooked Wenpo fruits must be shredded so that they can be blended with the cabbage evenly. A famous dish for the top Beijing restaurants in the 1920s and 1930s, it is easy to prepare and not expensive.

Ingredients include yellow heart of Chinese cabbage, fresh wenpo, sugar and crystal sugar.

原　　料
大白菜心200克，榅桲500克，糖20克，冰糖100克，清水30克。

制作方法
1. 将大白菜心切细丝备用。
2. 将鲜榅桲取籽，将清水和冰糖用锅烧开，冰糖融化成浓浓的糖汁，再倒入去完籽的榅桲，锅内温度达到35℃即可，倒入器皿中晾凉。
3. 将榅桲和切好的大白菜丝加糖拌匀即可。

备注：煮榅桲的火候要把握好，不能太软。

糯米红枣

红枣是传统的补血佳品，历来都受女士喜爱，红枣加糯米，又增加了它的口感弹性，口味甜糯，所以更受女士青睐。

Glutinous Rice Jujube

Rich in Vitamin C, glutinous rice jujube, or Nuomi Hongzao, is considered a typiine dish ordered for ladies.

Ingredients include kernel-free jujube, glutinous rice flour, honey, sweet olive and sugar.

原　　料

去核红枣200克，糯米粉200克，蜂蜜30克，糖桂花20克，白糖50克。

制作方法

1. 将红枣在温水中泡两个小时去核备用。
2. 将糯米粉加白糖，加入约90℃的白开水，快速烫熟搅匀成面团。
3. 将备用的糯米面团酿入红枣里面。锅中加水烧开，放入酿好的红枣，开锅后倒入不锈钢盆内。上笼蒸5分钟，取出摆盘，最后浇蜂蜜、糖桂花即可。

备注：枣要选果肉厚实的，糯米不能烫得太硬，否则会影响口感弹性。

江米镶藕

莲藕生食可润燥止渴，熟食可利水排毒，可荤可素，是非常受欢迎的菜品原料。江米镶藕香甜软糯，是典型的清宫廷菜，很适合老人和儿童食用。

Lotus Root with Glutinous Rice

A healthy cold dish, lotus root with glutinous rice, or Jiangmi Ou, tastes soft and tender, and is particularly good for spring. The lotus root for this dish is desirably of 7cm in diameter.

Ingredients include fresh lotus root, glutinous rice, brown sugar, sweet olive and honey. The glutinous rice shall be washed three times and dried before it is filled into the pores of lotus root with cooked brown sugar, honey and sweet olive, then boiled and steamed.

原　料

鲜藕1500克，糯米150克，红糖100克，蜂蜜50克，糖桂花25克，红枣30克。

制作方法

1. 选择直径约7厘米的鲜藕两节，放在水里用毛刷洗干净，去皮。
2. 糯米用凉水淘洗一下，放在通风处吹晾备用。
3. 将鲜藕一端切去一小口，沥净藕孔里的水分，切口朝上竖起，向藕孔里灌入晾好的糯米，灌满后把切口与原来的那块藕片用牙签插在一起，防止糯米漏出。
4. 锅内倒入凉水，加入红糖、蜂蜜、红枣和糖桂花，烧开后放入插好的糯米藕，改小火煮3～4小时，待藕变成紫色时，捞出晾凉，切成0.5厘米厚的片，放入盘中，浇上煮藕的原汁即可。

备注：不能用铁锅，否则藕会变成黑色。

京糕梨丝

山楂糕原本的酸甜，加上梨的香甜，口味清新爽口，开胃健脾，止咳平喘。

Hawthorn Cake with Pear Strips

Blending hawthorn cake's sour-sweet flavor with juicy pear's fragrance, hawthorn cake with pear strips, Jinggao Lisi, is refreshing and tasty. It is best to serve in summer.

Ingredients include pears, hawthorn cake and sugar.

原　　料

梨两个约350克，山楂糕100克，白糖50克。

制作方法

1. 梨削去外皮，除去内核，用刀切成火柴棍一样粗细的丝，山楂糕也切成同样的丝备用。
2. 将切好的原料放入冰箱内冰镇20分钟，放入容器内，加入白糖拌匀，放入圆盘，盖上盖焖5分钟，开盖即香气扑鼻。

备注：一定要加盖焖，这样山楂糕和梨的味道才会融合而出特有的香气。

JE Mansion Cuisine *Cold Dishes*

健壹公館菜・冷菜

素什锦

将绿、红、黄等新鲜的蔬菜均匀搭配，是一道有益健康的绿色食品，口感清脆、爽口，色彩艳丽，给人一种更添鲜美的滋味。

Mixed Vegetables

An appetizer specially for summer, mixed vegetables, or Sushijin, is evenly assorted of fresh vegetables in green, red and yellow colors. Not only tasty, but it also looks pleasant and inviting.

Ingredients include fresh bean curd stick, shelled peanut, scallion, pricklyash peel, baby cucumber, broccoli, ginger, dried pepper, celery, soaked black fungus, star anise and sesame oil. The shelled peanut shall be soaked in cold water for 8 hours to prepare this dish.

原　　料

鲜腐竹20克，花生米25克，大葱10克，花椒10粒，小黄瓜25克，西兰花25克，姜10克，西芹25克，黑木耳15克，八角2颗，盐15克，香油10克。

制作方法

1. 将花生米用500克的凉水泡8小时备用。
2. 将鲜腐竹、小黄瓜洗净切成菱形块，西芹削皮切成菱形块，西兰花切成均匀大小，泡好的木耳用刀切两半，备用。
3. 将水烧开，加入泡好的花生米、盐、大葱、姜片、八角、花椒，煮5分钟捞出，自然晾凉即可。
4. 备一盆冰水，将切好的西兰花、黑木耳、西芹用开水焯15秒后捞出，放入冰水中冰凉备用。
5. 将花生米、西兰花、鲜腐竹、黑木耳、西芹、黄瓜块加入适量盐、葱油15克、香油10克拌匀即可。

备注：煮所有原料时一定不要煮过，否则口感不脆。

中学为体，西学为用：可以亲身体验的文化博物馆

健壹公馆
JE Mansion
中国·北京朝阳区道家园甲16号
Jia No. 16 Dao Jia Yuan, Chaoyang District,
Beijing, China
Tel: +(86 10) 5139 8739
www.china-je.com

生活在别处、家在旅途中：涤尽浮华的贵族园林

健壹景园

JE Mansion

中国·北京市海淀区玉泉山万安东路西洼果园18号

No.18 Orchard Xiwa, Wanan East Road

Yuquanshan, Haidian District, Beijing, China

Tel: +(86 10) 8259 6669 / 8259 6660

www.china-je.com

美轮美奂、尽善尽美的都市桃源

大宅门迎祥商务酒店
Auspicious Business Hotel
中国·北京市昌平区府学路23号
No.23 of Fuxue Road, Changping District, Beijing, China
Tel: +(86 10) 8011 6060
www.china-je.com

移步光阴变幻,坐看时空穿梭:美景、美食、美酒

上海科学会堂1号楼JE
Building 1 of Shanghai Science Hall JE
中国·上海市黄浦区南昌路47号科学会堂1号楼
Building 1 of Shanghai Science Hall, No. 47 of Nanchang Road,
Huangpu District, Shanghai, China
Tel: +(86 21) 3126 8801
www.china-je.com

健壹 公馆菜 JE Mansion Cuisine

图书在版编目（CIP）数据

健壹公馆菜 / 康健一主编. -- 北京：五洲传播出版社，2013.8

ISBN 978-7-5085-2576-1

Ⅰ.①健… Ⅱ.①康… Ⅲ.①菜谱—中国 Ⅳ.①TS972.182

中国版本图书馆CIP数据核字(2013)第181821号

主　　编：康健一
撰　　稿：熊　蕾
翻　　译：安仁良　Nick Angiers
图片提供：健壹集团
出 版 人：荆孝敏
责任编辑：王　莉
书籍设计：张亚静
印　　刷：北京雅昌彩色印刷有限公司

健壹 公馆菜

出版发行：五洲传播出版社
地　　址：北京市海淀区北三环中路31号生产力大楼7层
邮　　编：100088
发行电话：010-82001477　010-82007837
网　　址：www.cicc.org.cn
开　　本：230 × 280 1/16
印　　张：13
印　　次：2013年10月第1版 2013年10月第1次印刷
书　　号：ISBN 978-7-5085-2576-1
定　　价：380.00元